A Practical Guide to Biom

D1090289

Peter Agger • Robert S. Stephenson •
J. Michael Hasenkam

A Practical Guide to Biomedical Research

for the Aspiring Scientist

 Springer

Peter Agger
Department of Clinical Medicine
Aarhus University
Aarhus, Denmark

Department of Paediatrics and Adolescent
Medicine
Aarhus University Hospital, Skejby
Aarhus, Denmark

J. Michael Hasenkam
Department of Thoracic and Cardiovascular
Surgery and Department of Clinical
Medicine
Aarhus University Hospital, Skejby
Aarhus, Denmark

Department of Surgery
Witwatersrand University Hospital
Johannesburg, South Africa

Robert S. Stephenson
Department of Clinical Medicine
Aarhus University
Aarhus, Denmark

ISBN 978-3-319-63581-1   ISBN 978-3-319-63582-8   (eBook)
DOI 10.1007/978-3-319-63582-8

Library of Congress Control Number: 2017951216

© Springer International Publishing AG 2017
This work is subject to copyright. All rights are reserved by the Publisher, whether the whole or part of the material is concerned, specifically the rights of translation, reprinting, reuse of illustrations, recitation, broadcasting, reproduction on microfilms or in any other physical way, and transmission or information storage and retrieval, electronic adaptation, computer software, or by similar or dissimilar methodology now known or hereafter developed.
The use of general descriptive names, registered names, trademarks, service marks, etc. in this publication does not imply, even in the absence of a specific statement, that such names are exempt from the relevant protective laws and regulations and therefore free for general use.
The publisher, the authors and the editors are safe to assume that the advice and information in this book are believed to be true and accurate at the date of publication. Neither the publisher nor the authors or the editors give a warranty, express or implied, with respect to the material contained herein or for any errors or omissions that may have been made. The publisher remains neutral with regard to jurisdictional claims in published maps and institutional affiliations.

Printed on acid-free paper

This Springer imprint is published by Springer Nature
The registered company is Springer International Publishing AG
The registered company address is: Gewerbestrasse 11, 6330 Cham, Switzerland

Preface

So you are intrigued by scientific research and what it has to offer, but you do not know where to start? This book will help aspirational inexperienced researchers turn their intentions into actions, providing crucial guidance for successful entry into the field of biomedical research.

The world of science is exciting, and in contrast to what many people think, it is not confined to the intellectual elite, extraordinary genius minds or someone with a special gift. Science is something everyone can do. Like any other craft you just need the right tools, the right guidance and the right motivation.

Aimed at future researchers within the biomedical professions, be it undergraduate students, young doctors, nurses, physiotherapists or engineers, this book advises and supports novice researchers taking their first steps into the world of scientific research. Through practical tips and tricks we describe the entire research process from idea to publication, while also providing insight into the vast opportunities a research career can provide.

We hope that this book will help you make a smooth start in research, and aid and inspire you to create your own little piece of history in contributing something truly novel to the world of science. Who knows, you might make a career out of it!

Please help us improve this book for the benefit of future researchers. If you have any comments, questions or feedback, we would be happy to hear from you via guide.to.biomedical.research@gmail.com.

Aarhus, Denmark   Peter Agger
July 2017   Robert S. Stephenson
                                                        J. Michael Hasenkam

Acknowledgements

During the writing process of this book we have benefited immensely from the help of several key persons. First of all we would like to acknowledge the work and insight of the contributing authors, and the invaluable input of all our reviewers. Equally, we would like to thank Ken Peter Kragsfeldt for designing the figures in this book. All photos presented are provided by Shutterstock.com. The authors have obtained the relevant publication license for the use of these images.

Contents

Part I

Before You Start

The first part of this book will take you through all you need to know in order to prepare a scientific project. It will guide you through the different available research approaches and the process of finding a supervisor. You will subsequently find advise on how to define and describe your project.

1

The First Steps into Research

Throughout the history of mankind, curiosity and an open mind have been the driving force behind discovery. What is over that hill? What is across that body of water? What can be found in the dark depths of space? It was probably also curiosity that made you open this book. Essentially, research is about pursuing curiosity. Leaving no stone unturned, in an eternal search for new discoveries that will improve the understanding of ourselves, and the world around us. Imagine being the first person to describe a phenomenon that will forever change the way we look at ourselves. You could be just months away from producing something that lasts forever; a piece of science that will inspire and stimulate the curiosity of future researchers.

1.1 Why Do Research?

Grandiose introduction aside, there are countless reasons to get involved in research. On a personal level, maybe you wish to attain an academic degree or qualification, maybe you would like to boost your CV, or perhaps you just want to see what all the fuss is about? At this point you probably already know why you would like to get involved in research, that is the easy part, but what are the first steps you should take? Who should you talk to? Where do you go? How do you start?

Before you can answer these questions there are a couple of things you need to consider. First of all, which topic or field of research interests you the most? Genuine interest is arguably the most important driver for sustained engagement in research. A research project will take you through an entire

© Springer International Publishing AG 2017
P. Agger et al., *A Practical Guide to Biomedical Research*,
DOI 10.1007/978-3-319-63582-8_1

spectrum of emotions, from extreme happiness to deep frustration, and everywhere in between. An existing interest in your research field is not a prerequisite for success, but in times of trouble your passion for the subject is sometimes what gets you out of bed in the morning.

Second, you need to decide which type of research you would like to do. An important aim during your initial steps is to determine who you are as a researcher. How do you like to work?

Which approach is likely to suit you best? Are you a "lab rat" who likes vials and Petri dishes? Do you get your kicks from working with experimental animals? Are you intrigued by technical inventions and applications? Or would you prefer a quiet day at the office working with databases or questionnaires? Maybe you can only see yourself in the clinical setting, investigating fellow human beings? Obviously, no choice is scientifically better than the other, but some may be a better match with your personality and temperament. Just to make this decision a bit more complex, it is not uncommon to combine multiple approaches of research into a single project. It is important to consider all possibilities in the early phase, prior to choosing your supervisor and designing your study. In Chapter 2 we provide a thorough walkthrough of the different approaches to research.

"Anything that can go wrong will go wrong". In spite of the fairly negative connotations of Murphy's Law, it is indicative of one of the main challenges in research: the art of preventing things from going wrong. A skilled researcher is able to anticipate the obstacles and unexpected turns that are inevitably going to occur, and is able to navigate their project elegantly past them. This skill is based on one thing and one thing only: **preparation**.

Preparation is essential from the very first moment you consider starting a research career. The more you know about your preferences, as outlined above, the higher the likelihood of finding the right project, the right supervisor, and the right research group.

Source: Shutterstock

1.2   Initial Contact

There are many ways of establishing contact with a research environment. You may already know someone who works in a group you would like to join, or maybe you know a professor who is conducting research you are particularly interested in. Face-to-face contact is always the most effective way of instigating collaboration, but it is not always possible. An email to a potential supervisor may do the trick, but it might also be lost in the hundreds of other emails academics receive on a weekly basis. There are no golden rules to follow in this matter, except that the lack of a response is not the same as rejection. Sometimes, establishing initial contact requires equal measures of perseverance and patience. In Chapter 3 we elaborate further on what you need to consider when looking for a supervisor.

1.3   The Project Life Cycle

Your first project will allow you to journey through the entire life cycle of a project. In the remainder of this book we will often refer back to the concept of the "Project Life Cycle" when describing the time line from your initial research ideas to the publication of your study. The Project Life Cycle is outlined in Figure 1.1. In the following section you will be introduced to some of the vocabulary used to describe the different stages of the cycle—yes,

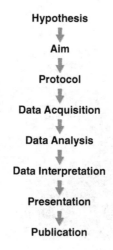

Fig. 1.1   Outline of the Project Life Cycle

scientists have their own special language—and it can even differ from one research area to another.

To enter the cycle you must first have a solid **hypothesis**. This is a kind of basic question or speculative statement specifically related to your research, it may be something like "I anticipate this new drug can cure cancer". Based on this hypothesis you formulate an **aim** for your study, e.g. "the aim of this study is to investigate whether drug X has a curative effect on cancer". Despite seeming to address the same issue, note the important linguistic transformation from the hypothesis to the aim. These and the other aspects of the Project Life Cycle are all put into context in your **protocol**, the description of your planned scientific project, which will be outlined in more detail in Chapter 8. The protocol is the cornerstone for planning and conducting your research project. It is used to convey to others what, why and how you will investigate your hypothesis, it is also the basis for applying for funding and the relevant approvals to conduct the study.

Once the protocol is ready and approved by all collaborators, including your supervisor, you are basically ready to go. Once all methods have been established, equipment is working, etc., you start **data acquisition**. At this point is it important to adhere strictly to the protocol. Do not make any changes once the protocol is closed and data collection has been initiated! This is one of the fundamental rules in science. If you change your procedure for data collection, you have effectively started a new project. Therefore, preparation at the protocol stage is extremely important—and pays off in the long run.

When all data has been collected according to the description in the protocol, you start the **data analysis** phase (Chapter 12). Here you perform all calculations, statistics, and graphic presentations, again, in accordance with what you defined in the protocol.

Then comes one of the prime times in the whole research process: **data interpretation**. What does the data actually tell you? Can your hypothesis be accepted or rejected? Remember, either outcome is scientifically valid and interesting. You must, therefore, analyse your data with an open mind and be objective. In this phase, it is important not to be **biased** towards a desired outcome. It is only human to hope your new drug works, but you must make all efforts to assess your data objectively. You are the first person in the world to see these data. Consider it a privilege to be the first to interpret them. Your co-authors can comment on your interpretation and suggest modifications, but you have the privilege of setting the scene.

Another important stage in the process is the **publication** of your findings. Here you shall demonstrate your ability, in collaboration with your co-authors, to disseminate your results to the wider community. This publication can be in the form of a **poster** or an **oral presentation** at a conference or a **scientific manuscript**. The highest-ranking publications are those, which have undergone a **peer review** process. We provide separate chapters on the various related topics later in the book.

No scientific process has been completed until the outcome of the study has been published. The scientific publication is both the hallmark and the formal end point for the Project Life Cycle. It is also a product, which is important in measuring the quality of your scientific work. The stronger the message, the higher the impact on the scientific community. At later stages the quality is measured further by the number of times other scientists cite or refer to your publication—more about this in Chapter 18.

Once you have completed a project life cycle for the first time, you have earned your stripes. You can now consider yourself a successful scientist in the making! You are now an acknowledged and skilled individual. You have demonstrated your ability to conduct a research project, which has gained the interest of the scientific community, and has conveyed an important scientific message. You are now at a stage where you can provide the answers to the questions you posed at the beginning of the cycle.

1.4   Time is an Important Factor

It takes time to initiate a scientific project. This is extremely important to remember. As you will appreciate from reading this book, there are many things you need to consider before you can actually start a study. Depending on the expected extent of your project, initial contact around a year before you plan to start is not at all too soon. Try to imagine how long it will take to find a supervisor, define a project, write up your project description, apply for funding and approvals, recruit study subjects and so on. No matter how long you may think it takes, it takes twice as long—at least!

1.5   Knowledge is Important Too!

As outlined in the Project Life Cycle, the cornerstone of every research project is a hypothesis or a research question that needs to be answered. A good hypothesis is very important, it is the key to success. Even negative results cannot destroy a study with a good hypothesis. Coining the perfect idea for a research project is difficult and often requires extensive knowledge within a scientific field. As a newbie in research you are, therefore, most likely not in a position to define the optimal hypothesis for your project.

This highlights an intrinsic dilemma in project planning. During the initial phase of the project life cycle, you will experience a mismatch between the importance of each decision that needs to be made, and the knowledge you have on which to base these decisions (Figure 1.2). In the beginning of a project, decisions need to be taken that will define the project from that point onwards. Although your knowledge and insight are limited at this point, these decisions still need to be the right ones. Conversely, at the end of a project where your knowledge is extensive, the potential consequences and the importance of a specific decision becomes less pronounced. In short, you need to boost your knowledge as quickly as possible when starting a new project. Again the word "preparation" comes into mind. This can be done by reading papers and textbooks, and by discussing your project with your supervisor and collaborators. Guidance on how to effectively search the scientific literature and subsequently read a scientific paper is presented in Chapters 5 through 7. It is evident that even at this early stage of your project, you are deeply dependent on a supervisor that can provide the insights needed to conceive the perfect idea or hypothesis. This is why we designate an entire chapter for finding the right supervisor for you!

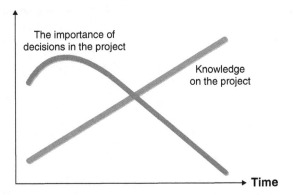

Fig. 1.2 The importance of individual decisions in research projects over time plotted together with the available knowledge. The latter can be improved by studying and supervision, but will always improve during the course of the project. Modified from Attrup and Olsson "Power i projekter & portefølje". 2nd edition. Jurist- og Økonomforbundets forlag 2008

1.6 Let's Get Started

In the remainder of the book we will walk you through everything you need to know to ensure a smooth journey through the Project Life Cycle. There are many things to consider, and in spite of thorough preparation, it is almost inevitable that you will forget some of them along the way. Do not worry about making mistakes. Rest assured that there will be time to correct mistakes, this book will, however, help you foresee most of the unexpected obstacles, and give you the extra energy to focus on what is really important …stimulating your curiosity.

2

Approaches to Research

In the first chapter, we asked you to consider what kind of researcher you are. But given the fact that you are most likely new in the field of research, how on earth should you know what kind of researcher you are? Well, deciding on which approaches to research suit you best is a good place to start. You do not necessarily have to decide on only one approach. You will find that projects often incorporate more than one approach, especially in the setting of multidisciplinary research. This chapter will provide an overview of the different research approaches involved in taking an initial idea all the way to its implementation as a new technique, treatment or ground-breaking theory. We then show how these approaches can be interconnected in a global view of research. You will then be in a position to make an informed decision as to which type of research activities are the best fit for you.

2.1   What are the Different Approaches?

Biomedical research has many faces and many fields of expertise exist in this discipline. In the following section, we walk you through the most common approaches. While reading this section try and identify which approach or approaches sound most appealing. This exercise will prove useful for future tasks such as finding a suitable supervisor (Chapter 3) and research group, and in defining, planning and writing your protocol (Chapters 4 and 8).

© Springer International Publishing AG 2017
P. Agger et al., *A Practical Guide to Biomedical Research*,
DOI 10.1007/978-3-319-63582-8_2

2.1.1 Computer Simulations

Very often, the first step when testing new ideas and hypotheses involves running computer simulations. Simply put, the simulations use a mathematical description, or model, to form a computer based dynamic analogue of the behaviour of something from the real world. Such investigations are often referred to as "in silico", and are typically low cost studies that do not involve patients or animals. A prerequisite for working with computer simulations is often a background in mathematics or computer science, thus it is unlikely that you as a biomedical researcher will be working first hand with this research approach. It is, however, very common to engage in collaborations involving computer simulations.

Source: Shutterstock

2.1.2 Laboratory Investigations

Most hypotheses in basic research are tested in the laboratory using cell lines, specimens or the like. This is often referred to as "in vitro" experimentation, referring to the notion that something is growing in a Petri dish. This setting allows you to investigate many very specific hypotheses, while controlling many extraneous variables, the human compatibility, however, is often low. It is, nevertheless, a crucial approach, as your hypothesis will often need to be validated "in vitro" before you can ethically justify an experiment in a living creature.

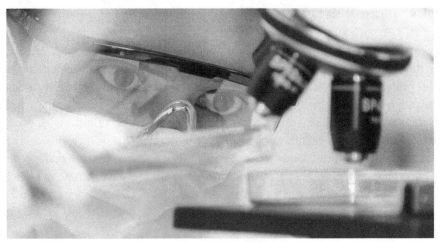

Source: Shutterstock

2.1.3 Animal Experiments

After successful passage of the controlled "in vitro" phase, the next level comprises a more complex biological investigation in experimental animals—to again test your hypothesis, but this time in a more realistic, and also more challenging setting. You now have less control over extraneous variables, but you are edging closer to a clinically relevant message. Animal experiments fall into two broad categories, "in vivo" and "ex vivo". Experiments conducted in living specimens, for example surgical interventions, are termed "in vivo", while experimentation on non-living tissue, for example genetic analysis of biopsies, is classified as "ex vivo".

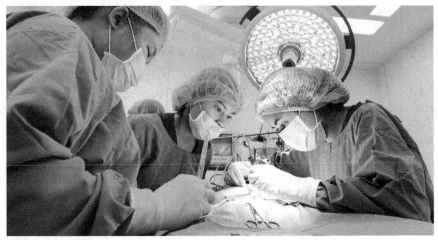

Source: Shutterstock

2.1.4   Clinical Research

If you do not find the prospect of working in the lab or with animals particularly appealing, you may find that working with actual human beings is your preferred research approach. Often the hypotheses investigated in clinical research have already been explored to some extent using lab-based approaches. As the name suggests, this approach is more likely to provide results with direct clinical implications within a single project life cycle. But bear in mind, such studies are often a product of multiple previous studies. There are many types of clinical research. Often you would think of a clinical experiment as a study comparing variables between a group of healthy volunteers and a group of patients with a disease, but here you should also consider registry-based studies as a type of clinical research.

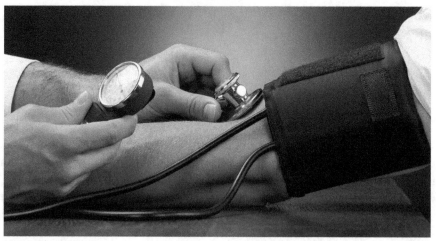

Source: Shutterstock

2.1.5   Registry-Based Studies

A very important side-branch of clinical research is the registry-based approach. Using various registries of archived data, you can extract information on a large number of human beings and compare numerous parameters to find associations between events and disease. The biggest difference from general clinical research is that you are working with very clinically relevant data. Using this approach you could potentially change the way patients are treated in a single Project Life Cycle. In this regard, registry-based studies can often feel very rewarding.

Source: Shutterstock

2.1.6 Clinical Trials

You can take clinical research to the next level by testing your hypotheses using a so-called clinical trial. Here you are not just comparing variables, you are testing an intervention. You enrol patients into a study where they are randomly selected to undergo either your new intervention, a placebo intervention, or the conventional intervention. Often clinical trials span several institutions and take several years to complete. Hence, it is highly unlikely that you will be conducting such a trial as a new researcher, but because they have a huge impact on how we treat patients, it is important to acknowledge their existence. They are often the product of many project life cycles, and can be seen as the ultimate goal for biomedical research. Some researchers can spend their entire career working towards such an approach.

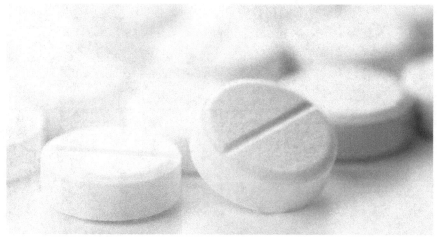

Source: Shutterstock

2.1.7 Meta-Analyses

Meta-analysis is a rigorous test of a research hypothesis. Here several studies, which tested the same hypothesis are grouped together in one statistical analysis, with the aim of testing the scientific strength of the research message. It is common for studies testing the same hypotheses to draw different conclusions. By grouping all of them together you can reveal the overall tendencies toward one conclusion or the other. This is particularly useful if it is difficult to gather large sample sizes in local communities. Like registry-based analysis, you are working with very clinically relevant data that can potentially change the way patients are treated.

2.2   A Global View of Research

In Chapter 1 we outline the concept of the Project Life Cycle. It is important, however, to appreciate the position of a single life cycle in the Global View of Research (Figure 2.1). Although your project may incorporate one, two or even three different research approaches, it is still, but a small cog in the Global View of Research. Many Project life Cycles, therefore, individually and in concert, contribute to the global view of research. Note, the various cycles do not come from you exclusively, they can derive from any scientist or research group in the world.

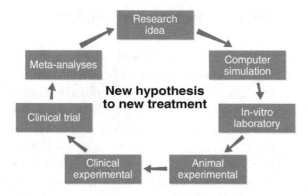

Fig. 2.1 Global overview of research showing the different approaches in the stepwise process from generation of a new research idea to implementation of a new treatment, technique or acceptance of new scientific fact

This is an important consideration when deciding on a particular approach. You have to appreciate that the impact of your work can often be dictated by your approach. For example, if it is your desire to be involved in the development of a new treatment, and your project implements an "in vitro" approach, you cannot expect the drug to be commercially available by the end of that particular Research Life Cycle. You will have to peddle a few more cycles before clinical application, namely animal experiments, followed by clinical experiments, and finally a successful clinical trial or two! The process from idea to daily clinical use can often take 10 to 15 years. One researcher will not be in a position to go all the way through the entire "Global view of research" (Figure 2.1). Realistically you may take a hands-on role in only one or two of the project life cycles. Research is a stepwise progression, where no single step can be left out and no step is more important than any other. Regardless of your approach you will be making a difference.

As a new researcher, you should aim to use the approaches, which make you feel most comfortable and through which you can contribute most. Even new hypotheses are conceived from existing knowledge. The global view of research is a continuum based on building upon existing knowledge, concepts and ideas to ultimately provide new and improved knowledge. This is not a new concept, it is viewed as a fundamental principle of science:

If I have seen further it is by standing on the shoulders of giants.
- Sir Isaac Newton 1675

2.3   The Right Research Environment

For your approaches to be successful you need to be in the right environment. Probably the most important factor in ensuring your scientific experience is a pleasure and not a burden is the presence of an inspiring research environment. Therefore you need to find a research group where you feel at home. It is always more fun and enriching to work as part of a research team. In this setting, you will gather research competences as a group, instead of having to learn everything from scratch. You will have the opportunity to develop unique skills, why not teach your fellow researchers to use those skills. You thereby contribute to raising the research competency of everyone in the group.

When you are about to enter a research group, allow yourself time to perceive the personal relations and the working atmosphere in the group. It is not something you can measure objectively or ask specifically about, you should just have an open mind and use your intuition. Very often, the nature of how you are welcomed in the group is a good indicator. Do you feel comfortable in the group? If so, you are probably in the right place! Most research groups have **Journal Club meetings**, where topics pertinent to the group are debated in depth. Such meetings can be a good forum for discussing and critiquing current scientific ideas and literature. The meeting coordinator may ask someone from the group to provide an overview of a certain topic, e.g. "What do we know about heart valve prostheses" or "What are the potentials and the characteristics of a certain analysis technique". Such meetings broaden your mind and give you an insight into areas of interest in your research field. It is a good idea, therefore, to get involved in group meetings as soon as possible, the knowledge gained will be invaluable when it comes to conducting your literature search (Chapter 5) and writing your protocol (Chapter 8). It is also a good way of socialising and getting to know the skills and strengths of your colleagues, so you can work more coherently as a group.

Like in any other group, social activities are also an important part of the fun. Do not underestimate the importance and benefits of joint sporting activities, cultural experiences, or even a simple drink or eating out as a group. The more fun you have in the group, the better you perform scientifically.

Intentionally, the above description of the research environment does not comprise discussion of internal competition, rivalry, distrust, scientific fraud and other negative perceptions some might have of how a research environment can function or malfunction. Although everyone should do their best to avoid such negative behaviour, when they arise, such issues should not be neglected. The strongest weapon to combat any negative behaviour in a research group is openness and freedom of discussion. By nature research is a competitive vocation, but it is certainly possible to compete while maintaining good social and professional relations.

3

The Right Supervisor for You

Your aim is to find a supervisor who is right for you. This chapter will advise you on how to identify, research and make that all important initial contact with your potential supervisors.

On the surface, finding a supervisor may appear a fairly trivial task, but do not take this decision too lightly. This is effectively the first scientific collaboration you will make, and like any effective collaboration you have to be able to work productively with one another. You will meet with this person on a regular basis, plan projects together, conduct experiments together and write scientific papers and grant proposals together; it therefore helps if you actually get along—at least to some extent! Beyond day to day professional responsibilities, your supervisor will also be a mentor, provide pastoral support and importantly be someone you can confide in when times get tough, which they inevitably will at some point. Furthermore, this individual will likely be a long-term presence in your career, providing a wealth of professional opportunities and access to new and exciting research networks. You can find out more about how to develop a successful research network in Chapter 19: The Scientific Network.

It may be starting to become apparent, that when finding the right supervisor, there are many factors, which should influence your decision. But note, you are not looking for the perfect supervisor; you are looking for the <u>right</u> supervisor <u>for you</u>.

© Springer International Publishing AG 2017
P. Agger et al., *A Practical Guide to Biomedical Research*,
DOI 10.1007/978-3-319-63582-8_3

3.1   A Good Place to Start

3.1.1 Defining Your Areas of Interest

If you ask any researcher when are they most motivated and most productive, the prevailing answer will be along the lines of "when I am working on a project I am interested in……fascinated by……passionate about", the same key words will remerge time and time again. This may sound obvious, but if you take on a project you have no passion for, there is a high risk that the project will suffer and ultimately, most of all, you will suffer. So, try to find areas of research you are interested in and passionate about, this may be a specific area you have encountered during your education, or have come across in the scientific or even general literature. Do not worry, however, if you do not know the exact area you would like to work in, start broad, start by thinking about an area of medicine or a particular disease that interests you, or how about a specific biological system or organ that intrigues you? You will soon start to home in on your area of interest. Many institutional websites and resources advertising research positions will be organised along these lines, so you will soon begin to navigate towards resources allowing you to pinpoint a few areas of specific interest to you.

At this stage, also think about what type of research you would like to conduct. We have walked you through the various research approaches in Chapter 2, but in short, research can fall into three broad areas: basic science, clinical research and registry-based studies. This is an important consideration, because generally speaking a supervisor's expertise will usually fall into one of these areas. For instance, you should not approach a basic scientist suggesting they supervise your registry-based study!

3.1.2 Your Position in the Hierarchy of Research

Before you contact any potential supervisor, you should first stop and appreciate your position in the hierarchy of research (Figure 3.1). Since you are reading this guide you are likely relatively inexperienced when it comes to research. As you may expect you have to start at the bottom of the ladder, you are a research student. But do not let this demoralise you, we all have to start somewhere, even the very top professors were once in your position. Like most professions moving up the research ladder is a progressive, rewarding and even motivating process. Remember you are not on your own, there will be plenty of other

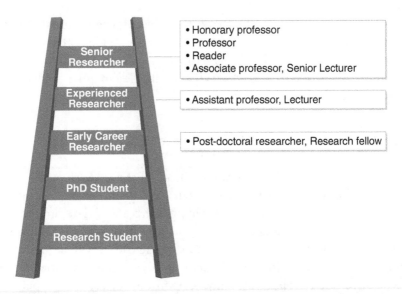

Fig. 3.1 The Research Ladder—Depicting the hierarchical structure of research

younger researchers either in your lab or the wider department—they can be a useful ally.

At the next rung on the ladder we have the PhD students, it is likely they will already have some research experience, be it a masters or undergraduate dissertation. The post-doctoral researchers and fellows are those who have completed their PhD. They are commonly referred to as "post-docs" and are found on the next rung of the ladder. They generally come under the umbrella term early career researchers, but will have substantial research experience. They are a great source of knowledge, and importantly still remember what it is like being a research student. Get what you can from them, be it insight into the field, the dynamics of the lab, or specific methodology tips, etc.

The experienced researcher is someone who has completed numerous post-doctoral positions or has acquired an academic lectureship position whereby research and teaching is now part of their role. It is these individuals who are the first individuals on the research ladder that may be a potential supervisor. The naming of such positions is quite confusing because the same type of position has different names in different countries, but is often an assistant professor or lecturer. They will be competent in grant writing and generating research projects. Finally we have the senior researcher, often the principal investigator (PI) of a major project or laboratory group, who has received national and international recognition, has vast experience in teaching, and is a regular generator of new research projects. In other words you are looking at a "Big Cheese" in research. Depending on the country or institution, they adorn titles such as associate professor, senior lecturer, reader, professor or even honorary professor if they hold a dedicated chair—the titles are many, and differ around the world. Assess what is behind the title in order to identify the right supervisor for you.

3.1.3 The Type of Project

Obtaining a research position at any position on the research ladder can be competitive, but do not let this cloud your judgement, do not just jump at the first project or supervisor who is willing to have you! The project has to be right for you, so thinking about the type of project you would like to be involved in is important. It is important to consider what type of project your potential supervisor can offer you. Remember this is a multi-factorial decision, the type of research project you embark upon can dictate what type of role you play within the supervisor's research group. For instance, you may wish to take on a pre-defined research project in which the majority of the scientific protocols and analysis methodologies are already developed and validated. In this setting you are almost certain to produce results of use, preventing the often demoralising prospect of conducting an entire study, which yields either inconclusive or invalid data. This can, however, be at the sacrifice of personal independence and intellectual freedom. If you envisage yourself as a more independent researcher, with a little more control over what and how the study is conducted, then perhaps a newly devised project is more appropriate for you. This approach has the potential to produce more novel data, with increased personal responsibility and satisfaction. But finding someone willing to take a chance on you, you are an unknown entity remember, may in some cases prove difficult. It is also very time consuming to develop a truly unique research project, and because of the uncertainty of the outcome, difficult to find funding for. As you can see there are pros and cons to both, however, there are no hard and fast rules, often the best projects encompass aspects of both.

You may also want to ask yourself whether you want to become part of a multidisciplinary team, the important journals look favourably on such studies, as do funders and research councils. But it is also important that you will have an opportunity to gain an identity within your group, and make significant contributions, which warrant primary authorship on publications.

3.2   The Type of Supervisor

Although personality will be a key contributor to the quality of the relationship you have with your supervisor, it is difficult to fully assess this in the brief encounters you are likely to have prior to committing to your research project. It is, therefore, important to consider their position in the hierarchy of research, and their current commitments.

A highly distinguished professor with lots of experience and influence in your chosen field can provide a wealth of knowledge and opportunities, which less senior academics often cannot provide. This could include access to novel methodologies, financial resources and influential collaborators. Inevitably, however, such an individual is likely to have limited free time, and may not be able to provide the same level of mentorship as a less senior academic, to some degree there is a trade-off. Try and assess the level of dependence you will place on your primary supervisor, there is no right or wrong amount, it is person specific. Just do not confuse initiative and independence with naivety. The level of contact time should be openly discussed with your supervisor; you should have clear mutual expectations as outlined below.

Co-supervision is common in most research projects, you may have a primary, secondary and even tertiary supervisor. You may benefit from having an experienced professor as a main supervisor, providing initial ideas and an international research network, and a younger researcher at PhD or post-doc level as a co-supervisor, providing day-to-day supervision. This way you can have the best of both worlds. In the setting of biomedical research it is to your advantage if you have both a basic scientist and clinician taking some role in your supervision. It provides a good balance of insight into your field and allows you to grasp the scientific and clinical implications of your work. This is, however, not always feasible.

3.3   Research Your Supervisor

The research starts here! At this point you have found a research area of interest, and identified a specific group or advertised position, and you would like to make a formal approach to the supervisor. To ensure the supervisor is right for you, and that you have a grounding in your prospective field of research, there are a few things you should explore before making that all important initial contact.

Start with the obvious things, where does the supervisor work? What academic position do they hold? And finally, and most importantly, take a look at their publication record, you can obtain a wealth of information from this exercise. If you are inexperienced when it comes to searching for publications see Chapter 5 "searching for scientific literature" for guidance. While researching their publication record, ask yourself the following questions; how often do they publish? Which journals do they publish in? What is the impact factor of those journals and is their work cited elsewhere? Answering these questions will give you a good idea of your supervisor's productivity, and the novelty and impact of their work. You will also gain insight into the level of activity in the field, perhaps try to think about future scope for your area. Furthermore, are they part of an extended research network, do they publish with other groups? Where do they fit in? Are they a first author, last author or somewhere in between?

Where might you fit in? Do they have a track record of providing opportunities for students to publish? You can easily assess this by looking at the author affiliation and qualification, if an author is from the same department, but does not have many letters after their name, specifically PhD, then they are likely to be lower down on the research ladder, and like you a student.

Finally, do not neglect other groups working on similar projects in the field. Generally speaking, most research groups or networks will have competitors. Knowledge of this will certainly impress your potential supervisor, perhaps even critically analyse their competitors work!

Source: Shutterstock

3.4   Meeting Your Potential Supervisor

Regardless of whether you are applying for an advertised position or simply enquiring about a potential project, your initial contact is important. An email is a good first approach. Make sure your email reads well and is free from typos and grammatical errors of any kind. To a supervisor a sloppy email means a sloppy approach to work in general. Make it short and concise. Busy researchers will not read long emails. Start by introducing yourself, concisely tell them about your research interests, and why you would like to work with them, but be specific, do not just flatter them! Also ensure the email reads as though it is targeted to them specifically, not just a generic email you may have sent to several other researchers. Give them an idea of your ambitions, but be realistic and focused. Suggest a meeting, but be flexible regarding your availability, the easier it is to set up an initial meeting the better. Remember academics and clinicians can be very busy and receive countless emails each day, so be patient. You may not get an instant reply, or even get a reply at all, this does not mean rejection, be resilient, adopt a five working days rule before re-approaching.

There are many resources available for interview, meeting and communication techniques, we will not explore them here. But during your meetings be enthusiastic and informed; let them know you will be hard working and fully committed to the project. Go in armed with your previous research of their current scientific activities, publication record and a general view of the field. You will not understand everything they say, they are an expert remember, so do not pretend you do. An intelligent insightful question will impress them more than a nodding dog! Feel free to propose new areas of study. The supervisor will be impressed if you are already thinking like a scientist, but be humble not demanding. A novice **proposing** an interesting new investigative approach is far more appealing than a novice **professing** to have the best ideas in the world.

3.4.1   Clear Mutual Expectations

Supervision is a relationship, which should facilitate good science and improve your skill set. Supervision is about feeling appreciated and accepted, and it is about being engaged, interested and inspired. This is true for both you and your supervisor. Obviously you need to find the right supervisor, but the supervisor also needs to find the right student. During your initial meetings with your supervisor it is a good practice to find out what they will expect from you during your time working with them, but this relationship is a two-way street, make sure they are aware of what you would like from them as a supervisor—have clear mutual expectations. Define these expectations from the start, this avoids problems down the road, do not get caught out by expectations you did not expect. Approach this issue diplomatically, and above all do not come across demanding. Many supervisors do not consider new researchers to be in a position to demand anything, but they will gladly engage in a friendly discussion on how to optimise your collaboration. Discuss how often you should meet, and set dates and times, be open and explain the level of mentorship you require. Discuss milestones and deliverables; ultimately as a researcher you are defined by your publications, is it a feasible expectation to publish a paper from your project? Discuss things like IT equipment and office space, it is important you feel like a member of the group. Discuss external opportunities such as visiting collaborating labs, grant writing and conference attendance. Finally, after your meeting take the opportunity to speak to the supervisor's current students and post-docs, their team will provide valuable insight into the labs working environment.

It is important to appreciate the commitment you make when you join a research group. You expect them to deliver and they expect you to deliver. Your supervisor has literally invested in you, they expect you to work hard and be committed to them. But you are not bound for life, you can always explore new fields in the future, this is perfectly normal. The research skills you acquire are universal, they are transferable to any field of research. But your reputation will follow you wherever you go, so make sure you maintain a good relationship with your supervisor. You never know when you might need them!

Once you have decided on the right supervisor for you, and importantly they have decided you are the right researcher for them, you have reached an important milestone in your journey for becoming a scientist; you have successfully formed your first scientific collaboration, congratulations!

4

Defining Your Project

So, you have secured your supervisor, now it is time to develop your project. Initially you need to decide whether the project is right for you, that is the easy part! There is then a wealth of practical and theoretical issues to consider when planning a project. It pays off to spend time planning your project in detail right from the start. This helps you foresee and avoid obstacles and pitfalls that could potentially delay your project. This chapter guides you through the considerations needed to define and plan your research project.

4.1 The First Considerations

Whenever you are considering a research idea for your future project, you should ask yourself these basic questions:

- Do you see a real need for the outcome of the research project? or is it "just interesting information"?
- What new information will the results actually tell us?
- Will it lead to publishable results?
- Do you have a realistic time frame to conduct the project?
- Do you have practical facilities to perform the project?
- Is the project just an opportunity for you to gain experience in research, or do you see it as a potential long-term investment?

Finally, and arguably most importantly,

- Are you genuinely interested in the research area? Does the prospect of conducting the project intrigue you?

© Springer International Publishing AG 2017
P. Agger et al., *A Practical Guide to Biomedical Research*,
DOI 10.1007/978-3-319-63582-8_4

Answering these questions will help you decide whether the project is right for you—or whether you should look for inspiration elsewhere. Your answers to the questions are not meant to be right or wrong. The purpose is to help you reflect on the suitability of the project, and help you identify areas that you may wish to change.

Remember science is a plastic, malleable and a forever evolving vocation. So, if there are aspects of your project you are not sure about, or wish to tinker with, speak to your supervisor. Transparent communication is a great skill to adopt. Your supervisor will be open to entertain intelligent suggestions for modifying the content or path of the project.

When evaluating potential alternative projects, the best place to start is to read scientific papers relevant to the research topic. Details on how to do this can be found in the forthcoming chapters. Do not be disheartened by the fact that many other scientists have already published results on your topic. It is actually encouraging—it means that, like you, other researchers also feel this research topic warrants investigation. Exploring "active research areas" can often provide you with the inspiration you require.

This is not to discourage you from aiming to develop novel concepts or explore a new or emerging research topic. This task can be challenging for a new researcher, but not impossible!

Source: Shutterstock

4.2   Practical Availability

Defining your project is mostly desk work, which can be performed anywhere, provided it is supplemented with meetings with your prospective supervisor. In the initial phase of planning your project, the most important person of reference is the supervisor. It is important, therefore, to get a strong sense of the availability of your supervisors for the entirety of your project. You will be very much reliant on them during the phase where you write your protocol, and also when you start performing experiments. When data collection is flowing nicely and the initial phases of data analysis are underway, presence of the supervisors is less critical. However, when it comes to data interpretation and the initial steps of publishing your results, then they become important again. To avoid getting stuck due to lack of availability of your supervisors, prior planning is essential.

When defining your project you should take all elements of the Project Life Cycle (see Chapter 1) into consideration. However, many other elements relate to the practical feasibility of conducting your research project.

Ask yourself; can the supervisor provide a research environment with other young researchers, a secretary, laboratory assistance, desk space? Access to IT, for example, a computer, printer and Internet access? What about equipment for performing experiments and analysis of data? Although not absolute necessities, these are nice to have; and it is at least good to know if they are not available. It is all a matter of aligning expectations. You should, therefore, not only ask yourself such questions, but also your supervisor.

Regardless of which research field you enter, in order to sculpt your project, you need the right set of research tools. The availability of such tools, be it, analysis software or laboratory equipment is critical for conducting your research. Starting a research project with only limited tools is not an ideal scenario. Often procurement of new instruments is part of conducting a research project, but you should at least have a time frame for having those instruments at your disposal. It can be very frustrating to have a great protocol for a novel study in a well-respected research group, but have no tools to work with. Avoiding such a situation relates very closely to good planning.

Another critical element when considering practical availabilities is money. Like in any other aspects of modern life, it is difficult to do anything without money. A well-established researcher, such as your supervisor, often has an economical buffer to work with, so you can start working before acquiring specific funding for your project. This makes life a lot easier!

Not everyone is that lucky. Sometimes part of your scientific training involves writing grant applications in order to acquire funding for your project. You must always be aware of the financial standing of your project, make sure you clarify any funding issues with your supervisor from the start. We guide you through the process of applying for funding in Chapters 8 and 10.

4.3   Time Management of a Research Project

As a rule of thumb, you can perform the majority of your research work within 10% of the allocated time. When planning any project, however, bare in mind it is the final parts, which will take up 90% of the time! (Figure 4.1). Very often new researchers are pressed for time at the end of a project, often due to poor planning during the early stages of their project.

When planning a 12 months project, for example, many supervisors will recommend outlining a project that you, as a new researcher, believe you can finish in 4 months. Knowing full well it will in fact take you 12! The main message here is to be realistic, do not put yourself under undue stress by setting yourself an unachievable time frame. If you do, both you and the project will suffer.

Setting a realistic time frame is, therefore, a critical part of planning and executing a research project. But at present you are too inexperienced to make an accurate assessment of the time needed to conduct each aspect of the

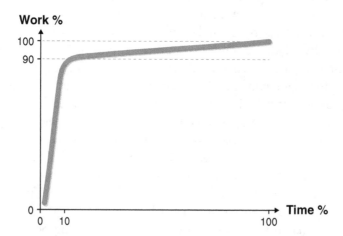

Fig. 4.1   The relationship between the time spent on a project versus the completion of the project. The first 90% of a project takes 10% of the allotted time, whereas the last 10% of the project takes the remaining 90% of the time

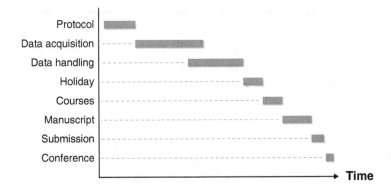

Fig. 4.2 The Gantt diagram is an excellent way of visualising the time frame of your project. Each element has its own line corresponding to its expected time frame. Some elements may be overlapping because two things can be handled simultaneously

Project Life Cycle. You should consult your supervisor and fellow researchers for specific advice on time management. There are, however, some tools you can further aid the process.

One of the best tools for research planning is the so-called Gantt diagram. As shown in Figure 4.2, you can easily assign a time scale to individual tasks within your project, and then compare the forecasted time scale with how things are actually progressing in real time. Such a time calibration is very informative, but it is also a very good reporting tool. Share your Gantt chart with your supervisors, then they can easily assess your progress, and pin point what resources, such as help, equipment, or even funding, you may need. The Gantt diagram is also a good tool for maintaining an overview of the project.

Another important exercise in time planning is to counteract delays by inserting time buffers. In reality, everything takes longer than you expect. The more experienced you become, the more realistic you become, and the better you will be at planning and executing a project.

4.4 Redefining Yourself

An interesting bi-product of conducting a research project is your change in mental state, owing much to the level of accountability and responsibility that comes with it. For most of your life you will have been a student, presented with the task of learning from whatever is put in front of you, be it textbooks, lectures or study guides. In order to carry out a successful research project, however, it is essential you take responsibility and adopt the role of project coordinator. You are no longer learning from the discoveries of others. Now you are in unknown territory, with the task of discovering new information of your own. You are the director responsible, in part, for all elements of your project, successful completion of your project is predominantly down to you! You very often also adopt the role of team leader, since you have to integrate your work with that of your collaborators, and make sure everyone performs their allocated task in an ordered and timely manner. The independence associated with research is attractive for some people, but the notion of responsibility can be rather intimidating for others.

Through the course of the project, you must remember to keep your supervisor informed. During your project, you will in many ways become more knowledgeable than your supervisor. Thus, it is easy to run ahead and lose him or her along the way. Do not make this mistake, your supervisor can provide invaluable guidance and insight at the end of your project. If you have not kept your supervisor up to pace during the project, it is difficult for him or her to help you.

In science, the transformation from novice to expert is very short. You can become part of the international elite very quickly. This is why it is very motivating to work in research—you will very quickly obtain recognition for the work you do.

Part II

Conducting Your Research

This part of the book will guide you through the relevant skills and considerations required when conducting a scientific project.

5

Searching for Scientific Literature

Contributing author:
Karen Tølbøl Sigaard
Librarian
Master of Library and Information Science
Aarhus University Library, Denmark

Searching the literature is an integral part of doing research; it allows you to identify all existing knowledge in your new research field. Literature can be found in a number of ways, for example through colleagues, quick PubMed searches or news feeds. These are all legitimate ways of finding information; however, when you start a research project you want to have a more complete picture of what has been written on a topic. For that you will need to use a more systematic approach. This chapter focuses on how to systematically search for the essential literature you need to support your research project.

5.1 Introduction

When you begin a literature search, you should be aware of three things. Firstly, literature searching takes time. You may be eager to move on to other parts of your research project, but the time invested in doing a systematic literature search is well spent. It can help you avoid repeating research that other scientists have already done, whilst also enabling you to know how your research relates to that of other researchers in the area. Secondly, a literature search is often a very iterative process, where you test different search term

© Springer International Publishing AG 2017
P. Agger et al., *A Practical Guide to Biomedical Research*,
DOI 10.1007/978-3-319-63582-8_5

combinations, evaluate results, and go back and add, change or remove terms. Descriptions of the literature search process often make a literature search seem more linear than it is, because this is the simplest way to present the search process. Thirdly, the goal of a literature search is not a perfect search strategy, rather the acceptable balance between resource expenditure and expectations of the search result. As such do not expect your literature search to disclose everything that has been published in your research area.

The effectiveness of a literature search can be evaluated using two concepts: precision and recall. **Precision** indicates how successful your search is to find only relevant references, while filtering out irrelevant references. **Recall** is the ability of your search to find all relevant references. It can be useful to think of a search as a balance between recall and precision. A search with high precision will be focused and produce only a few references, with high relevance, but there is an increased risk of missing other relevant references. A search with high recall is very comprehensive and therefore the risk of missing relevant publications is lower; however, the search result will most likely be larger and include many irrelevant papers.

Before starting the main literature search you can do an initial search to identify any existing reviews. Time and effort can be saved if someone else has already done some of the work of finding and critically appraising relevant primary studies. Furthermore, if a systematic review already exists, you can use its search documentation to identify potentially relevant search terms.

5.2   Developing a Systematic Literature Search

This section focuses on the basic techniques of literature searching and how to build a systematic literature search strategy. It also introduces the most important biomedical literature databases.

5.2.1 Search Techniques

You need to master some basic search techniques to search the bibliographic databases. These techniques help you tell the database how you expect it to interpret your search query. They are applicable in all bibliographic databases, although the symbols used to represent a given technique can vary from database to database. Check the help section of the database you are using to see the available symbols.

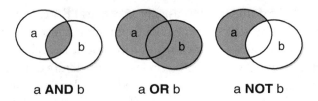

Fig. 5.1 You can combine search terms to narrow or widen your search by the use of boolean operators such as AND, OR and NOT. "AND" and "OR" are used most commonly in searches, whereas "NOT" is used much more sporadically. This is because you risk filtering out relevant publications

The most important technique is the use of the **boolean operators** AND, OR and NOT. They often need to be written in capital letters to distinguish them from search terms. The boolean operators control how the database combines the search terms when searching for more than one term. "AND" indicates that both words must be present in the references of the search result. Using "OR" means that all references will be found that contain either or both search terms. The "NOT" operator tells the database to retrieve references that contain the first but not the second word. The effect of the three operators are visualised in Figure 5.1.

5.2.1.1 Phrase Searching

Phrase searching can be used to indicate that two or more search terms should be searched as a phrase. Usually, quotation marks toggle phrase searching. If phrase searching is not used, and you have written no boolean operator between two words, databases will normally assume that you want to search the words with an implicit "AND", and therefore retrieve all references containing both words regardless of their relative positions. For example, searching for breast cancer will find all references in which the two words are represented, whereas searching for "breast cancer" will find only the references that have the two words standing next to each other in the given order.

5.2.1.2 Truncation

Truncation is a technique by which you can search for different endings of a word. When using truncation, the database searches for all the different words it can find that start with the characters you have indicated. This is useful when searching for both singular and plural or both the adjective and noun form of a word. The symbol for truncation is usually an asterisk. For example, searching schizophreni* will find schizophrenia, schizophrenias, schizophrenic, etc. You should, however, be cautious when using truncation, though, as you might get unexpected results if your search term is not sufficiently unique. Searching for kid* will include kid and kids as expected, but also irrelevant terms like kidney and kidnapping. Some databases allow for a variation of truncation called wildcards. Wildcards are symbols, often ? or #, that can be placed inside a word, substituting zero or one character in case there are several ways of spelling a search term. For example, p?ediatric will look for both pediatric and paediatric.

5.2.1.3   Proximity Operators

When combining two words with a proximity operator you indicate that both words have to be present in the references retrieved and that they must be within a given proximity of each other. The proximity is indicated as x number of words that can at most be between the two search terms. Databases do not agree on how to write the proximity operator, but it will most often consist of a letter or word and a number; for example, searching for "randomised NEAR/5 trial" in the database Embase will find references in which "randomised" and "trial" are no more than five words apart, such as "randomised controlled trial" or "randomised double-blinded trial".

5.2.1.4   Parentheses

More than one search technique can be used in one search. For example, "post-partum haemorrhage" AND prevention will look for references containing the phrase "postpartum haemorrhage" and the word "prevention". Searches also often consist of more than two words. When you search for multiple terms using both AND and OR you must tell the database in which order you want the search query to be resolved. If you do not, the database will use its default priority setting, often meaning that AND-searches will be performed before OR-searches. To control the order of the search you can place the different parts in parentheses. For example, (therapy OR treatment) AND (anxiety OR ocd OR phobia) will search for references containing at least one word from the first parenthesis and at least one word from the second parenthesis.

5.2.2 Translating a Research Question into a Search Strategy

Your literature search is defined by the scientific question you want to answer, written in your protocol as the hypothesis and the aim of your project. The question is your compass when deciding exactly what to search for and how to do it. This is important to remember throughout the search process since it is easy to get sidetracked by articles about other interesting aspects of your topic. The more precisely you can formulate your question, the more efficiently you can execute your literature search. For instance, it is much harder to construct a useful search from the question "I want to know something about hypoxia-ischemia" than from "what studies exist on using rat models of hypoxia-ischemia in neonates?".

Once you have defined your research question you can formulate a search strategy. A search strategy outlines what words you intend to include in your search, how you plan to combine them and what limits you are going to use if any at all. This is the overall framework for your literature search and it will help keep it concise and manageable. Building a search strategy is partially a trial-and-error process so be prepared to alter the specifics of your search strategy when you start searching to allow for changing, adding or removing search terms when needed. The final version of the search will be documented in your search protocol as outlined in the Section 6.4 on "Search Documentation".

In order to find your search terms, you must identify the most essential parts (key concepts) of your question. These key concepts are aspects of your search that can be combined to form a searchable version of your question. If your question consists of several key concepts you might need to decide which to include in your search. Here you should take into consideration which concepts are most likely to appear in titles and abstracts of potentially relevant articles, due to the fact that you cannot search the full text of the article. The remaining key concepts can instead form part of your inclusion/exclusion criteria; that is the criteria by which you sort the references identified in your search. In the above example, the key concepts would be "hypoxia-ischemia", "neonates" and "rats". When you have your key concepts, you can start gathering different synonyms for them.

A prerequisite for a successful search is that you identify all, or the most important, synonym terms for your key concepts, i.e. "neonates" and "new-born". There is no definitive method by which you can tell that you have the right words, but there are different methods for identifying relevant terms. Firstly, you can draw on your own and your colleagues' knowledge of the area. Secondly, you can get input from already known articles or books, which you have read that cover different aspects of your question. Thirdly, you can make initial, explorative searches and get inspiration from titles, abstracts and keywords, etc. The aim is to find the different terms authors of relevant articles have used to describe your concept. You can also consider including search terms of differing abstraction levels, e.g. rats and Sprague Dawley. When you have a list of different synonyms for your key concepts this list should be expanded with relevant inflections and spelling variations. Consider singular and plural word variations as well as American and British spellings. For example, if you search for "hypoxia-ischemia" you can also search for "hypoxia-ischaemia" and if you search for "rat" you can also search for "rats".

5.2.3 Block Search

When organising your search terms, you can think of your search as a collection of building blocks that need to be combined. In one block, you put the first key concept and all its synonyms, inflections and spelling variations. In the next block, you put the second key concept, and so forth. All terms within a block are combined with OR, to find all references containing at least one of these terms. The blocks are then combined with AND. This search retrieves all references containing at least one term from each block, in all possible combinations.

The block search method ensures that your search is structured and it helps you keep focus throughout your search. Using a block search table from the beginning can help you to maintain an overview of the process and guide you from question to search strategy. In the example of searching for studies on using rat models of hypoxia-ischemia in neonates, your block search table could look like Figure 5.2.

Apart from the search terms the search can also include one or more limits, for example publication year. If you were doing the search in Figure 5.2 you could choose to limit it to publications written in English. Furthermore, if you had a systematic review from 5 years before your search covering the topic you could also decide to only search for publications published in the last 5 years.

When you have defined your general search strategy, this strategy must be adapted to fit the features of each of the databases you are going to search. In practice, the general search strategy is developed using the first database, and subsequently transferred for use in the other databases. Note that some of the terms in Figure 5.2 have the suffix [MeSH], which is PubMed specific; see Section 6.1 on "Controlled Subject Headings". When running the search

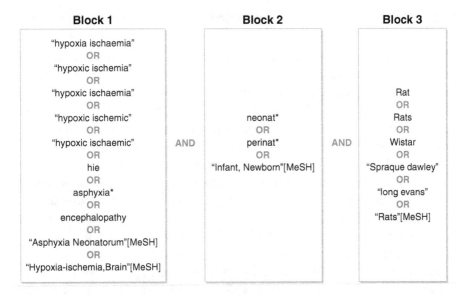

Fig. 5.2 Structuring your search into blocks of synonymous terms ensures that you will find more relevant papers in one search

strategy in the first database, it is important to evaluate whether the search result is what you expected. You will never get a search result where all identified articles are 100% relevant, but scanning the titles on the first result page will tell you if you are retrieving what you intended without too many irrelevant articles. If you already have one or more relevant articles, or know of relevant authors, you can check if they are found in your search. If not, you probably need to alter something.

It is not possible to construct a search strategy that guarantees finding all relevant publications. Making a thorough and systematic search strategy, searching multiple databases and using supplemental search methods makes it less likely that you miss something relevant, but also make your literature search more time consuming. In the end, you must find the balance between time usage and thoroughness most appropriate for the purpose of your search. Constructing a solid search strategy and implementing it in a structured way will optimise the balance between the time you decide to spend on your search and the quality of the outcome.

5.3   Reference Management

The search results can be sorted directly in the database immediately after performing the search, if your search is simple, your search result is small and you are only searching one database. If the search covers multiple databases and yields large numbers of references, the sorting process can be facilitated more efficiently by using reference management software. Reference management software is basically a piece of software that organises all your references in one place, regardless of their format (book, article, web page, report, etc.). References can be imported into the software directly from the databases, or created manually. Furthermore, the software can help you identify duplicates in your search results, which can be removed before you sort the references. The imported references can be organised in folders, PDFs with the full publication can be attached, and in some cases you can read and annotate your references within the software. All these functions can help you manage your references.

Reference management software is also useful when writing a manuscript for your scientific publication (see Chapter 16). It can communicate with your text editing software and format both in-text citations and reference list in the preferred reference output style. For example, if you are writing your references in the Vancouver style, which is a number-based citation system, the reference management software keeps track of what reference has which number, and if you remove a reference from your text it changes the numbers and updates the content of the reference list. This will save you a lot of time and tedious work.

There are numerous different types of reference management software on the market. Some of them are free while others require a license to use. The available software changes continuously and, therefore, any recommendations on which specific software to choose would quickly be outdated and is thus not included in this book. Different software has different strengths and weaknesses and hence it depends on your preferences and needs, which programme would be best suited. Discuss with your supervisor or your colleagues in your group, which reference managers they prefer. If you follow their preferences they will be able to provide guidance and tips and tricks to get you started.

5.4   Databases

Bibliographic databases contain references to scientific publications. Most databases require a license to access and search, PubMed being a well-known exception. Access to a bibliographic database is not the same as access to the full text of all the references indexed in it; full text access is a separate issue, which will not be dealt with here. A bibliographic database holds a record of each of the publications indexed in it. For example, a record of an article contains, when available, author names, article title, journal title, volume, issue, pagination, abstract, author keywords, author affiliations and controlled subject heading. Apart from the controlled subject headings, which will be discussed in another section, the different types of information about each publication are searchable, either individually or all at once. Each category of information is called a field (e.g. the author field) and searching all, or a subset of, fields is called free text searching.

Which databases are relevant to search in a given situation is mainly dependent on the topic, but also on the extent of the search. If the goal is to find a small number of highly relevant articles, it will probably be sufficient to search the most relevant database. If the aim is to uncover almost all publications on a topic, you will need to search multiple databases, even if they overlap in content coverage, to heighten the chance of finding all potentially relevant references. The following list contains some of the most important databases in the health sciences. The databases are presented in very general terms. Since they are highly dynamic, a detailed description would quickly become outdated.

PubMed: International database, developed and curated by The National Library of Medicine in USA, covering all topics in the biomedical research area. PubMed can be searched free of charge.

Embase: International database maintained by the Dutch publishing company Elsevier. Covers more European journals than PubMed, but is in many other aspects comparable to it. Embase requires a license to access.

CINAHL: The Cumulative Index for Nursing and Allied Health Literature indexes references to articles, book chapters, dissertations, etc., focusing on nursing literature. CINAHL requires a license to access.

PsycInfo: Database covering literature with a psychological focus. PsycInfo requires a license to access.

Cochrane Library: Is a collection of several individual databases, the most important of which are The Cochrane Database of Systematic Reviews (CDSR) and The Cochrane Central Register of Controlled Trials (CENTRAL). Cochrane Library requires a license to access; countries can have a national license.

Scopus: Large database covering all research fields, although the health and natural sciences are better represented than the social sciences and the humanities. Scopus indexes citation data and can therefore be used for citation searching. Searching Scopus requires a license.

Web of Science: Large database covering the same research fields as Scopus. Like Scopus, Web of Science is a citation database. Web of Science requires a license to access.

This list is in no way exhaustive. Even more specialised databases exist, but most likely the above will fulfil your needs. If not, contact your local librarian. In this chapter, you have been introduced to the thoughts and techniques underlying the literature search process. As with any research skill, you must practice it to become proficient. If you need more detailed guidance on, for example, the use of a specific database, contact your local research library or just read on to the next chapter on advanced literature search.

6

Advanced Literature Search

Contributing author:
Karen Tølbøl Sigaard
Librarian
Master of Library and Information Science
Aarhus University Library, Denmark

Knowing the literature is paramount in science. There is no need to repeat what others have already done and no excuse for not citing the most ground-breaking papers in your field. Hence you may find that you need some more advanced tools to make sure you have found all the relevant papers. In the following chapter, different ways of expanding on and refining your literature search will be discussed. This chapter will also cover the process of documenting the search and handling the outcome.

If you are just reading this book front to back to get a quick overview of the process lying ahead of you, you may find the contents of this chapter a bit dry and slightly irrelevant. But we assure you it is not! Here you will learn a lot that can make your life in research much easier in the long run. It is, however, completely acceptable to skip to the next chapter for now and return when you actually need to do a literature search. Then the tips and tricks provided in the following will make much more sense to you and be a lot easier to remember.

© Springer International Publishing AG 2017
P. Agger et al., *A Practical Guide to Biomedical Research*,
DOI 10.1007/978-3-319-63582-8_6

6.1   Controlled Subject Headings

In many of the biomedical databases, the records are enriched with searchable controlled subject headings as a supplement to free text searching. Controlled subject headings are keywords predefined by the database. When an article is entered into the database, the appropriate headings are assigned to the record based on the topic of the article. Using controlled subject headings when searching has several advantages. Firstly, it adjusts for synonyms. All articles about children will get the heading "child" regardless of whether the author wrote child, children, childhood, schoolchildren, paediatric, etc. Secondly, a record will only get assigned a subject heading if it is about that subject. When searching free text for the term "mindfulness" the search will also retrieve a record with the phrase "studies of patients receiving mindfulness as treatment were not included in this review" in the abstract. Searching "mindfulness" as a subject heading will retrieve articles that are actually about mindfulness in some way. Thirdly, subject headings are organised hierarchically. In PubMed, "Sepsis" is a subterm of "Systemic Inflammatory Response Syndrome", which is a subterm of "Inflammation". This is useful when you want to search a whole category. If, for example, your question concerns all patients suffering from mental disorders, you can search the heading "mental disorders". The different types of mental disorders are subterms to "Mental disorders" and therefore automatically included in the search. If the same search was to be done in free text search only, it would be necessary to explicitly list all the different psychiatric diseases.

Not all databases have controlled subject headings, and it varies from database to database which concepts are available. When they are available, the database has a thesaurus where you can check if a term exists as a controlled subject heading. In PubMed they are called Medical Subject Headings or MeSH-terms; in Embase they are Emtree words. The individual subject headings for a publication are shown as part of the record; therefore, a good alternative method for identifying relevant headings is to browse those of relevant articles. A focused search can consist solely of subject headings, but often a thorough search will include both that and free text terms.

6.2   Broadening or Narrowing Your Search

If your search is yielding too many results, you may wish to make it more focused. One way to do that is to choose more specific search terms. Let us say, for example, that you are looking for studies on treatment of patients with pneumonia, but you are really only interested in pneumonia contracted while admitted to a hospital. Instead of just searching for pneumonia you could consider searching specifically for hospital-acquired pneumonia to narrow the search. Alternatively, you can restrict your controlled subject heading to major terms, which are the headings covering the main topic of the article. Another way to narrow the search is to remove search terms, especially free text words, from one or more of your blocks. Sometimes a term introduces too much noise to be useful, even if it could in theory find relevant articles. If you search the words one at a time when building your block search, you can easily identify which terms retrieve high numbers of relevant results. A third way of narrowing a search is to add another block. This can only be done, however, if there is a meaningful aspect to add.

If, on the other hand, your search is yielding very few results, you should, first, evaluate if this is what you expected. If not, you may not be using the right search terms. In that case, you need to figure out which terms you are missing and add them to your search. If your search terms are appropriate, you can consider widening the search. You can do this in several ways; as one would expect, these ways are more or less the opposite of the ones used for narrowing the search. One way is to remove a block from the search. Choose the least important block or the one least likely to be consistently reported in a record's title, abstract or controlled subject heading. Another way is to add more synonyms to your search to make it more exhaustive. This makes it less likely for references to be missed because the author used an alternative word for the concept you are searching for. A third way is to add more general terms for your key concepts to your search block. This way you will catch articles discussing your topic on a less specific level.

Source: Shutterstock

6.3   Supplementary Searching

Searching the bibliographic literature databases in a structured manner using keywords and subject headings is the main part of a systematic literature search. This search method can be supplemented by one or more other search methods.

6.3.1 Grey Literature

Grey literature is literature that is not yet published, and maybe never will be. An example of this is descriptions of ongoing clinical trials. It is also literature in formats that generally do not get formally published, like reports or guidelines. An often-mentioned reason for looking for information on research that has not yet been published is to avoid publication bias stemming from the fact that trials reporting positive or significant results are more likely to get published than trials reporting negative or neutral results. Guidelines, reports, etc. contain information that you may not be able to find in the published literature. The challenge, when searching for grey literature, is that it is not collected in well-structured databases like the published literature. Instead, it can often only be found on the website of the organisation from where it originates. However, some repositories for specific kinds of grey literature do exist. If you are looking for in-process trials, websites like *clinicaltrials.gov*, *opengrey.eu* or the International Clinical Trials Registry Platform could be relevant, even if they do not cover all trials. *Prospero* registers records of in-process systematic reviews. Searching for this type of to-be-published grey literature is mostly undertaken when the literature search has to be very comprehensive, for example if it is the basis of a systematic review.

6.3.2 Citation Searching

Citation searching identifies potentially relevant publications by means of the links created between publications through their reference lists. The point of origin is one or more relevant articles. It is assumed that publications cited by these articles or publications that cite the articles are in one way or another semantically linked. When authors cite a publication, they give credit to the research presented in it because they have used it in some way. This is a very important step in the research life cycle. Be aware, though, that citations can be both positive and negative, either crediting or discrediting another publication.

The semantic link between publications created by citations means that if an article is highly relevant to your topic, then at least some of the references that its authors chose to include are likely to be relevant as well. Searching the citations given by a publication is straightforward since they are always listed on the publication's reference list as part of the article. Searching the citations received by a publication requires the aid of a citation database. As mentioned above, Web of Science and Scopus are the two main citation databases. Citation searching takes place later in the search process, as relevant articles first need to be identified. Citation searching can both be used to find more relevant articles and to check if the topic search was thorough enough, in which case no, or only few, relevant additional publications should appear in this search. How much time, if any, should be spent on citation searching depends on how much priority you give to thoroughness in comparison to time consumption. Checking reference lists is usually done as an integrated part of reading relevant publications whereas citation searching requires more time.

6.3.3 Hand Searching

Hand searching is the process of manually browsing through content lists of publications, for example the newest issues of a specific journal or the contents of conferences proceedings. Like citation searching hand searching allows you to potentially catch publications missed in the topic search. Hand searching, like searching the grey literature, has the potential to discover studies of references to material not indexed in the bibliographic databases. It can be very time consuming, and, therefore, it is mostly done as part of a very comprehensive literature search.

6.4 Search Documentation

We strongly advise you to log your steps throughout the search process to keep track of what has been tried and what worked. The final search strategy should be documented in a search protocol. This is because a literature search is a research method and just like any other research method it needs to be documented to make the process transparent, evaluable and repeatable. Incomplete or no search documentation makes it difficult or impossible to know how you ended up with the references you have. This applies both if you are required to formally report your search strategy as part of your methodology and also if the documentation is only to be seen by yourself, your supervisor or your colleagues. Furthermore, a research project can easily take so long that a search update is required to make sure nothing new has been published in the meantime. Most databases offer the possibility of saving searches for later use.

The purpose, when writing your search protocol, is to make the documentation so precise that your searches could be repeated from this information. The following information is essential:

- **Databases searched**

  When documenting the databases, both the database and search interface should be recorded. For example, when you search PubMed you search a database called Medline through the interface PubMed. Medline can also be searched via an interface called Ovid. The interface defines which search techniques are available, and, in the case of PubMed, also to some extent the content of the database.
- **Date of search**

  Databases are dynamic; new references are constantly added, controlled subject headings are revised and information on existing references is updated. You can, therefore, get different results from performing the same search at two different points in time, even if the searches are only a few weeks apart.
- **Search terms**

  All terms, both free text and controlled subject headings, are essential to document. This documentation needs to be very precise to be able to repeat the search. For example, it should be noted if and how truncation, quotation marks, proximity operators, etc. were used, and if free text terms were searched in all fields or only in a subset (e.g. title and abstract). The search terms used should be noted for each individual database, since search technique and controlled subject headings vary.
- **Combination and limitation of search terms**

  It should be noted how the search terms were combined using "AND" and "OR" or parentheses. Any limits applied to the search should also be recorded.
- **Other means of searching**

  If any supplementary search method is utilised, this should be described as well.

The above are the most important details to record about the search. Reporting a literature search in such a way that it is reproducible requires a certain level

Search protocol

Medline (PubMed), searched 01.05.2017

Search ((((("Psychotic Disorder"[Mesh]) AND "first episode")) OR first episode psychos*)) AND (("Risperidone"[Mesh]) OR risperidone[Title/Abstract]) Filters: English	135
Search ((((("Psychotic Disorder"[Mesh]) AND "first episode")) OR first episode psychos*)) AND (("Risperidone"[Mesh]) OR risperidone[Title/Abstract])	141
Search ("Risperidone"[Mesh]) OR risperidone[Title/Abstract]	8879
Search risperidone[Title/Abstract]	7868
Search "Risperidone"[Mesh]	5600
Search (("Psychotic Disorder"[Mesh]) AND "first episode")) OR first episode psychos*	2936
Search first episode psychos*	2228
Search "Psychotic Disorder"[Mesh]) AND "first episode"	2241

Embase (Elsevier), searched 01.05.2017

No.	Query	Results
#12	#4 AND #10 AND [english]/lim	523
#11	#4 AND #10	561
#10	#6 OR #9	13129
#9	"risperidone"/exp/mj	7006
#6	risperidone:ab,ti	11998
#4	#2 OR #3	8915
#3	"first episode psychos*"	4246
#2	"psychosis"/exp AND "first episode"	8852

Fig. 6.1 Suggestions for tables of search documentation. *Top panel* shows a table generated from Medline (Pubmed). *Bottom panel* shows similar table from Embase

of detail, which is often underestimated. Even systematic reviews often lack in their documentation of the literature search. Figure 6.1 shows an example of a search protocol in which all the information listed above is present.

7

How to Read a Scientific Publication

The sheer number of publications within the scientific literature is enormous. Chapter 5 will help you hone in on the publications most relevant to your project. This chapter will help you to read the results of your search quickly and efficiently, thereby allowing you to survey a large amount of scientific literature and focus further on those really interesting publications. This is a great skill to acquire, and it is really not that difficult to learn, trust us!

7.1   Reading Scientific Papers

"Life can only be understood backwards; but it must be lived forwards". The words of the Danish philosopher Søren Kierkegaard apply not only to life, but also to scientific literature. As we will describe in Chapter 16 the scientific paper is written following a specific sequence of sections. You can, however, often benefit from reading it in a different order. Take a look at Figure 7.1 to get an overview.

When you read a paper your quest is to gain an overview and isolate important information fast. Start out by reading the title page. Does the title provide any clue of the study purpose, design or results? Proceed quickly to the abstract and spend some time here to get a more complete overview of the research message and main results. Figure out the specific aims of the study, maybe you need to skip to the end of the introduction to get that information. Then spend some time looking at the figures. Do they provide any answers to the study aims and hypotheses? Then starts the actual reading. Start with the introduction to get a complete overview of the rationale of the

© Springer International Publishing AG 2017
P. Agger et al., *A Practical Guide to Biomedical Research*,
DOI 10.1007/978-3-319-63582-8_7

When WRITING	When READING
• Title page	• Title page
• Abstract	• Abstract
• Introduction	• Aims
• Aims	• Introduction
• Materials and Methods	• Conclusion
• Results	• Discussion
• Discussion	• Materials and Methods
• Conclusion	• Results
• Acknowledgements	• Acknowledgements
• Disclosures	• Disclosures

Fig. 7.1 The differences between writing and reading a paper. The sections should be read in a different order, to allow for optimal understanding of the paper in a time efficient manner

study. Subsequently skip to the conclusion, normally found at the end of the "Discussion", to see what the authors think is the most important message from their study. Then look for the thoughts behind the conclusion and in particular the relations to other studies by reading the discussion. In principle you can stop here. You now have a good idea of the described study and its conclusions.

For the papers most relevant to your study, you may want to know the specific details of their methodology. Is it appropriate? The quality of a study lies in the correct use of scientific methodologies. Therefore, if you want to read further into the paper you need to read the "Materials and Methods" section. If the paper is really good you can read through the results as well, but you will find that most of the important details will be presented in the figures. Hence, you only need to read the remainder of the paper if you have a particular interest in the minor details.

7.2   How to Read a Scientific Paper in 2 Minutes

After you have completed a literature search, it is highly likely that you will find yourself with countless potentially important papers that you wish to read, but the prospect of reading through each and every one may seem unrealistic.

Luckily, you can quickly obtain a good impression of a paper by asking yourself a few simple questions:

- Does the title convey to you a clear research message?
- Where do the authors come from? Are they from a renowned institution?
- What is the hypothesis and the aim?
- What is the study group size and type? (in vitro, animals, patients)
- What do the figures tell you?
- What is the conclusion and does it answer the aim of the study?

Force yourself to find answers to the questions above as quickly as possible. You should be able to do it in about 5 min. With a little practice you will soon start shaving time off your personal best. Before you know it you will be reading papers in 2 min flat!

After having a reasonable overview of the subject, you might want to go deeper into specific aspects of your selected topic. For example, you might want to search for a certain measurement technique. To conduct this search it is essential you know exactly where in the scientific manuscript you are most likely to find this information. In this case, you should look in the "Materials and Methods" section. Take all the papers you found appropriate using the 2-min reading technique, and focus directly on which measuring technique the different authors have used. Again, you do not need to read the entire article to find this information.

Using the approach described above, you can search for information on many other aspects such as specific hypotheses or aims, and make a record of any relevant information by annotating and categorising the relevant references. Initially this "quick-read" approach may feel a little rude. After all, authors spend a lot of time and effort publishing a paper, but it is a necessary process if you wish to efficiently identify the most essential papers, which warrant a little more of your time.

You will not be able to remember many of the papers you read in this way. Therefore, it is important to make brief notes as you read. This avoids wasting energy re-reading them at a later date. Devise an appropriate system for organising your notes—do not rely on your memory, it will quickly become overloaded.

7.3   To Print or Not to Print?

If you are the type of person who likes everything printed out on paper, and like to make handwritten notes in the margins—get rid of that habit! Train yourself to read papers on the screen; make digital notes and use referencing software to handle the vast amount of information you will quickly accumulate. On top of doing the environment a favour, you are doing yourself an even bigger favour, because your literature collection will be searchable on your computer, as will any notes you make. For this task the reference management software we mentioned in Chapter 5 on literature search comes in handy.

7.4   How to Read Reviews

The best way for you to get insight into a certain research area is to start with some review papers. But note, the above approaches do not hold true for the reading of reviews. The authors of such papers have often already made the task easy for you by gathering the essential information on for instance a disease, a research methodology or even the relation between the two. You can get an overview of a review, of course, by reading its abstract. Also you can scroll through the paper focusing on the different subsections. Maybe you find some sections more interesting and relevant than others, but often you need to read a review from beginning to end to extract all the relevant information. Another advantage of review papers is the list of references, it gives you a comprehensive overview of all the relevant literature written on the subject. These references are a great resource to guide further reading on specific issues within the subject of interest. Here you have a gold mine of additional information, start digging!

8

The Scientific Protocol

The first document you will have to write when developing a new project is the scientific protocol. In some research environments it is referred to as the project description, because this is essentially what it is; a description of your research question and a detailed plan on how you are going to answer it. The scientific protocol for a single research project will consist of approximately 5–10 printed pages. Larger projects with many elements or multiple research questions may of course lead to more extensive protocols, but in general a protocol is a concise document. Writing a scientific protocol for the first time is a challenging task that requires a lot of preparation, and benefits greatly from regular interaction with your supervisor. It is, however, absolutely a worthwhile task. A well thought out scientific protocol saves you time, it helps you anticipate and control for potential pit falls or problems which may occur during the life cycle of the project. This chapter gives you the overview of what a protocol should comprise and how you design it.

8.1 Purpose of the Protocol

The protocol should convince the reader that the project is well thought out, novel and seeks to answer an important research question. It should also identify the weaknesses of your project and describe appropriate countermeasures. As we have briefly touched upon in Chapter 4, you will need this document to support funding applications and your enrolment at your institution. In short, it is the key to getting your project off the ground. The protocol should convey the message of a high quality study taking place in a top class research

© Springer International Publishing AG 2017
P. Agger et al., *A Practical Guide to Biomedical Research*,
DOI 10.1007/978-3-319-63582-8_8

environment. The quality of the study is often dictated by its research question, and the proficiency of the methodology proposed to answer it. The protocol should, therefore, convey a reasoned and appropriate research design using rigorous and feasible methodologies. Lastly, you should convince the reader that the team behind the study is indeed qualified to complete the task.

It is of course important that the protocol is well written and has an appealing appearance. A good protocol has an informative and concise title, a pertinent background section that contains clear research questions, a realistic and well-considered budget and timetable, and is generally clear and well ordered with a logical layout.

8.2 The Target Group

Before you start writing your protocol you must consider its target group. This is because different target groups speak different languages and want different things. Normally a scientific protocol is targeted at scientists within your field, but not necessarily within the same specific research area. If you are, for example, working in biomedical research, your protocol should be aimed at medical researchers and doctors, and if you are working within the field of engineering, your target group should be fellow engineers. The easier the protocol is to understand, the more likely you are to convince the reader that your project is of high scientific quality, and that you are the right team for the task.

You should bear in mind that the protocol serves different purposes depending on where you send it. Universities generally focus on the overall feasibility and scientific quality of the study. The same applies to foundations and awarding bodies, but additionally they also like to see that the budget is well defined and reasonable. This is in contrast to, for instance, an ethics committee. They have less interest in the financial aspects of your study, and tend to focus more on the ethical issues and safety aspects for animals and patients. But like the foundations, they tend to care less about the methodological technicalities. You should, therefore, consider the different target groups and write your protocol accordingly. Sometimes you may need to have different versions of your protocol, each modified to fit the requirements of a specific target group. Be aware that once your protocol has received official approval from either the ethics committees or the animal expectorate, any changes you wish to make may require separate approval, and in turn, revision of your protocol.

8.3 The Scientific Language

Up until the mid-seventeenth century, Latin was the language of science. Today the language of science is **English**. From now on, regardless of your mother tongue, most of your scientific communications will have to take place in English. Obviously, you should follow the standard procedure in your group, but in general it is advisable to write your protocol in English. This will train your brain to think in the right terms and sentence constructs. Effective communication of a scientific message, in English, is essential. You are going to have to face it, from now on every communication relating to your research will mostly be in English. So why not start as you mean to go on, after all, practice makes perfect!

The quality of your scientific protocol is very much based on your communicative skills, and your ability to produce a document, which is appealing to read. Try putting yourself in the shoes of the reader, ask yourself, is the project convincing? Is the message compelling? Would you fund it? Make sure that you **inform the reader; do not try to impress** them by displaying everything you know about the subject. A protocol should be a pleasure to read—not a burden!

8.4 Writing the Protocol: Structure and Content

The protocol should contain the following sections:

- Title page
- Objectives and impact
- Introduction
- Materials and methods
- Time table
- Resources and feasibility
- Ethical concerns
- Budget
- Publication strategy
- Perspective

As you can see, a protocol is a rather extensive document, but in the following we will guide you through every single aspect of it.

Overall your language should be concise, to-the-point and without superfluous information. Try to avoid the use of abbreviations. They impede the flow and can be exceedingly annoying if you are not familiar with them. Be very careful when deeming an abbreviation as common knowledge, often they are not, and hence they only serve to disturb the reader. This applies to any written document in science, not only the protocol.

> *Abbreviations help only the author, NOT the reader!*
> - Prof. Robert H. Anderson

8.4.1 The Title Page

The first page of the protocol is the cover page, see Figure 8.1. The cover page should contain the title of the study, your name as the project leader and your affiliation. A neat illustration can be added to make the protocol a bit more inviting. The title should be informative, brief and clear, while at the same time interesting and catchy. After all, what makes you pick up a book from the bookshelf?—Very often it is the title. Coining the perfect title is far from easy and it often requires multiple attempts to get it right. The best titles are usually simple and direct, while conveying the research message. Avoid trendy titles and word play. Try to use action-oriented verbs like develop, produce, prevent, reduce, etc. in the title.

8.4.2 Objectives and Impact

This initial paragraph can be added to further attract the attention of the reader. Present your research problem, project objectives and expected impact. This should appear as the first paragraph on the first page of your project description. You should also highlight what is new and unique about the project. It may also be relevant to argue why it is relevant to carry out the project now and not in the future. This could be because of political priority, cohort availability, newly developed techniques, or use of new breakthroughs, etc.

When describing the objectives, you must define the problem you want to solve. It is a good idea to describe the problem in relation to societal impact and describe the scope of the problem.

PhD Protocol

Myocardial Remodelling due to Right Ventricular Dilatation in Congenital Heart Disease

---

Peter Agger, MD

AARHUS
UNIVERSITY
DEPARTMENT OF CLINICAL MEDICINE
CARDIOVASCULAR RESEARCH UNIT

Fig. 8.1   Apart from the title, the title page should contain information on the author and relevant institutions. Moreover, you can benefit from a graphical element relevant to the project

When describing the impact, focus on the effects of your project results. Be concrete, ambitious and realistic. The following questions can guide you when writing this paragraph:

- Why is it important to carry out the project?
- What results do you expect from the project?
- What is the expected outcome of the project?
- What becomes possible when the project results are applied in practice?
- What effect will the project have on a given target group or society in general?
- Who has an interest in the project and why?

8.4.3   The Introduction

The background or introduction section is where you build the case for your study. It should provide a thorough and up-to-date introduction to your research topic. It is not the intention to cite every published paper with relevance for your topic, but you should go through the literature and find the best and most relevant references. You rarely need more than 10 references in a protocol. Often it is a good idea to adhere to original research and systematic reviews. You should start out by presenting the broad clinical perspectives of the study and then narrow down to your specific area of interest. Then proceed to identify gaps in the current knowledge and describe what has already been done to fill these gaps. If possible, also describe your own group's contributions to the area. Wrap up by stating what is still missing in literature, leading directly to your hypothesis and from there introduce the aim of your study. The writing process of the background section is a constant narrowing of focus. Figure 8.2 illustrates this narrowing concept.

As touched upon in Chapter 1, the hypothesis is the main research question that you would like to investigate. It should be clear and concise, and crucially it must be measurable. It must be relevant and novel. A study can comprise several hypotheses, but limit the number to three or fewer if possible. The final element of the introduction is the aim of the study. The aim is an overall broad statement consolidating the hypotheses into one sentence, which encompasses the measurable goal of the project. You can have multiple hypotheses, but there can be only one aim.

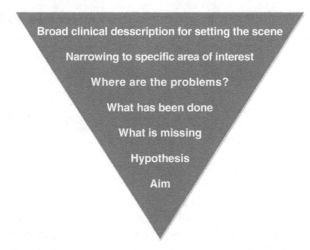

Fig. 8.2 The writing of an introduction section. Start with a broad clinically relevant description of the overall setting. Then narrow down to your specific area of interest. Define what are the problems and what has already been done. Outline what is currently missing and use that as a spring board to state your hypothesis and finally your aim

Example
We hypothesise the following:

1. Drug X can cure cancer.
2. Patients have fewer side effects from drug X compared with drug Y.

Thus this study aims to investigate the curative effect and potential side effects of drug X in the treatment of cancer.

8.4.4 Materials and Methods

This section is the cookbook. Here you describe in detail how you intend to test your hypothesis. In principle, another skilled researcher should be able to conduct your study using only the materials and methods section as a reference. You can build up this section in many different ways depending on your type of study, but in general you should consider including the following subsections:

- Study Design
- Study Population
- Experimental Procedures
- Data Handling and Statistical Considerations

8.4.4.1 Study Design

What type of study are you planning? Is it a clinical trial, an animal experimental study or something else? Describe how you will compare your study subjects. They could for instance be compared with a group of healthy controls, or a control group subjected to placebo treatment. Have you taken any precautions to avoid systematic errors? Perhaps by the use of blinding or randomisation?

8.4.4.2 Study Subjects

Describe who or what you intend to investigate. Is it humans, animals or cells, and if so what species or type? Are they modified in any way? Is it in vivo or ex vivo? Here it is important to state how many subjects you plan to include in the study, where they are coming from and also state how you intend to recruit them. If your study subjects will be assigned to groups, you should state this, along with a description of how they will be allocated. Finally, any inclusion or exclusion criteria should be clearly described. This mainly applies to human studies and could comprise parameters such as age, gender and specific diseases.

So how do you decide on how many study subjects you need? The more subjects you have, the higher the chance of finding a significant difference, but you also have to take feasibility into account. Sometimes it is simply not possible to include more than a limited number of subjects owing to financial or time related issues. If you want to know how many subjects you need to find a difference of a certain magnitude, you can conduct a **sample size estimation**. Here, you take the experiences of previous studies and combine them with your desired probability of finding a significant difference. This is a mathematical procedure, and there are designated applications available to calculate sample size estimations. We encourage you to investigate this matter further on the Internet if you need it, as we will not elaborate further in this book.

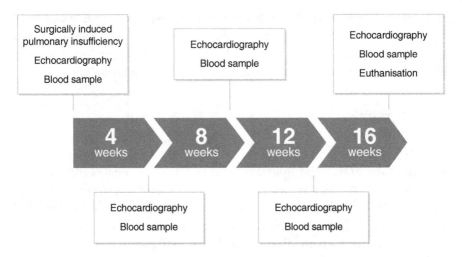

Fig. 8.3 An example of a study design. The study is spanning 16 weeks with different interventions every fourth week

8.4.4.3 Experimental Procedures

In this subsection you should describe all the techniques you will use to gather your data. In short, what do you intend to measure and exactly how and when are you planning to do it? Every methodology should be described in detail. Sometimes a flow diagram is an illustrative way of describing when you plan to do what, see Figure 8.3.

8.4.4.4 Data Handling and Statistical Considerations

In this section you state how you will sample, store and handle your data. Briefly state which parameters you will measure and how will you analyse and present them?

Being relatively new in the scientific environment you are not expected to be able to master medical statistics; hence this section largely relies on input from your supervisor. Many institutions have designated statisticians you can consult, which can be a very advantageous acquaintance at the planning stage of a scientific project. The aim of this section is to display that you can devise a reasonable plan for statistical analysis of your results. Data can be statistically interpreted in a number of ways. Statistical tests will provide results regardless of whether they are theoretically valid. This introduces the opportunity to force your results in a desirable direction. You should obviously avoid this. It is important, therefore, to show that you have devised a reasonable and valid protocol for data analysis. Provide a small paragraph on your considerations regarding data analyses and statistics. We dig a little deeper into the most important statistical considerations in Chapter 12 on Data Analysis.

8.4.5 Timetable

Lastly, you should show that you have considered when you should conduct each part of the study. It is important to be realistic. No one will support your project if they think your agenda is too tight. Whether you provide a time line or a timetable, it should convey the message in an illustrative way. In Figure 8.4 we show an example of a time line, this is just one way of illustrating the time course of a study. A Gantt diagram as shown in Figure 4.2 in Chapter 4 is another popular way of outlining the time frame of a scientific study.

8.4.6 Resources and Feasibility

It is obvious that you are never going to make it through the project on your own. You may well need technical assistance, or you may need help with data acquisition. Now is the time to describe the research group. Name each individual and mention his or her specific role in the project. Having a strong team of competent people can increase your chance of success tremendously, so remember that your supervisors also belong in this section. It is also in this section where you solidify your role as the scientific project leader—the coordinating and driving force behind the project.

If you need any special equipment or machinery, you need to explicitly state where that equipment is, and that it is available to you. This should convince

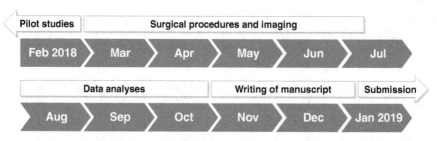

Fig. 8.4   An example of a study time line

the reader that it is feasible to conduct your study at your institution and that you are the right researcher to do it.

8.4.7 Ethical Concerns

A paragraph on the ethical concerns is mandatory. Ethics must be considered in all studies and not only if you are working with living creatures. In this paragraph you must delineate how you are planning to deal with these issues. There are rules and regulations that you must indicate you have considered. These issues are discussed in Chapter 9 on legislation and ethics. You should state that you have applied for, or intent to apply for, all the relevant approvals before initiating the project. It is worthwhile to discuss this with your supervisor at an early stage. First of all, because there are huge variations in the approvals needed for a specific kind of project, and second, because actually getting the approvals can often take quite some time. That said, the first step is to complete your protocol, because it is a basic requirement in most applications for approval.

8.4.8 Budget

Any protocol should include an overview of the financial costs of the project, and should contain a budget preferably in the form of a table. In smaller studies the entire budget may fit into the protocol itself, but in larger applications containing several projects, an abbreviated budget is usually inserted, and the full budget is then submitted as an appendix. Remember to include everything in the budget because foundations rarely cover expenses that were not stated in the initial protocol. It is obvious that running costs and salaries are to be included, but often researchers forget to include expenses for conferences and publications. You are most likely looking forward to the prospect of travelling around the world presenting your work, but conferencing is expensive, so that must also be included in the budget. As must publication expenses. Publishing papers is not always free of charge. High impact journals often charge a considerable amount of money for publishing your work. Figure 8.5 shows an example of a budget. Note that in the figure 30% is added to the subtotal labelled 'Overhead for the institution'. Some institutions require that you add this overhead, which is to cover administrative costs. Ask your supervisor whether this applies to your budget.

Surgical procedures incl. pilot study	#	Price	Total EURO
Animals, 5 kg.	25	118	2950
Surgical equipment and anaesthetics	25	479	12225
Animal caretaking 60 days @ 50 EURO	25	402	10050
MR-scanner usage	25	918	22950
Chemicals (formalin and PBS)	25	40	1000
Salaries			
Project leader	12	1300	15600
Technical personnel incl. MR-technician	25	297	7425
Other expences			
International conferences	1		1300
Publication of manuscript	1		2000
Subtotal			75500
Overhead for the institution		+ 30%	22650
Total			**98150**

Fig. 8.5   An example of a project budget

8.4.9   Publication Strategy

It is important that your results become publicly accessible, and adding just a small passage on your publications strategy shows that you have considered the entire project life cycle. Mention if you are intending to present your work at an international scientific conference. If you already know which one, state that as well. You should state whether you intend to publish both negative and positive results in an international peer-reviewed journal. Most likely you do not know exactly to which journal you will end up submitting your work, because that may depend significantly on your results, thus it is perfectly understandable to omit such specific information. As well as formal scientific publications, channels of communication could include newspapers and even social media platforms. Be as specific as possible. Think of the most effective way to reach the target group in question.

8.4.10   Perspectives

The perspectives section is one of the last sections in the protocol. This is where you elaborate on the potential impact of your study. Ask yourself how the new knowledge generated by your study will benefit patients, provide new scientific tools or be of societal value. At this stage the experienced reader will go back to your hypothesis and your aim and assess whether your perspective is directly connected to your stated aim.

Not everything will turn out the way you planned it, but in this section you should outline the best-case scenario, where everything goes according to plan. Reviewers will quickly recognise if you expect too much, therefore, be ambitious but do not exaggerate.

8.5   Assessing the Quality of the Study and of the Protocol

When the evaluation committee of, for instance, a university research committee or a foundation board, assess your protocol, they will mainly evaluate the quality of the proposed study. A high-quality study has a pertinent research question or hypothesis and an appropriate study design. They will assess whether your methods are appropriate and feasible, and whether your research team is sufficiently experienced. A good protocol is well structured and reads well. Spend time working on the layout, you want your protocol to look welcoming and intriguing, and remember to proof read your protocol several times to make sure you have corrected all typos and linguistic mistakes. Ask a colleague not directly involved in the protocol to proof read it if possible.

8.6   Who Should Write the Protocol?

Some supervisors may be tempted to write the initial protocol themselves. In that way they ensure that the protocol is proficient and it saves them a lot of time and effort supervising a young colleague. Tempting as it may seem, this is not an advisable approach. We always encourage new researchers to write their own protocol. Nothing gives you a better overview of the forthcoming task than preparing a protocol. In our view new researchers should always play a dominant role in developing their scientific protocols. During the process you will gain an extensive knowledge of your new scientific field; knowledge you will frequently call upon and implement throughout the life cycle of your project. Additionally, ideas and relevant questions often emerge from the naïve mind of a new researcher. You should, therefore, embrace the academic freedom that comes with writing your own research protocol, and demonstrate your ability to be the driving force behind a research project.

9

Ethics and Commercialisation

In any game, knowing the rules is the first step towards victory. The same applies to research, know the rules and play by them! If you do not pay attention to the rules and laws applicable to your scientific study, you risk running into problems that will slow the progression of your study, or end it entirely! It is important that you spend energy on this matter early in the project life cycle. You do not want to make important decisions based on insufficient knowledge of legislation and ethics that cannot be changed later on. This chapter will take you further into the ethics and legislations that apply to any scientific project, and help make sure that you do not overlook important matters before proceeding with the actual task of conducting your study.

9.1 Always Consider Ethics

In nearly every kind of research you need to consider the associated ethical issues. Ethics relates to the part of scientific philosophy dealing with what is right and wrong. In research this question can essentially be reduced to "Can an individual suffer in order to make life better for others?" and if that is the case, how much can we allow others to suffer? That has been defined in the Helsinki Declaration, which is the international constitution for ethical behaviour in science. It is being updated regularly since ethical behaviour today is not what it was 20 years ago.

© Springer International Publishing AG 2017
P. Agger et al., *A Practical Guide to Biomedical Research*,
DOI 10.1007/978-3-319-63582-8_9

Nearly all kinds of studies require some sort of ethical approval, either from an ethical committee or, in the case of animal studies, from an Animal Experiments Inspectorate. Some institutions have a local human ethics committee that has the authority to approve studies within a certain ethical complexity. In other situations ethical approval for research is passed on a regional or national level.

Ethical approval is very important, and in some cases it has a major impact on your study design. You should never start a scientific study until you have all the relevant approvals, not only is it unlawful, but any data acquired in this way is unpublishable! We can in no way provide a complete overview of ethics in general in this book, but we assure you that you will receive a lot of questions regarding this matter throughout the entire project life cycle. Furthermore, acquiring ethical approval can, on occasion, be a long drawn out process, just ask any experienced researcher! So it is advisable to get the ball rolling as soon as possible.

9.2   Know the Law

In every country there are laws on how you can conduct research, how you can use experimental animals, how you can access public databases and of course how you may use humans as study subjects. You may also find legislation on how you should store your data in various situations. The laws relevant to research differ tremendously between countries and because this book should be applicable to as many countries as possible it is very difficult to be specific in this section. In the following we will point towards some areas of research where it is important to consider legislation. If you obtain any approvals or permissions for your study, we suggest keeping a folder with all your permissions and remember to renew them if needed.

9.3   Working with Dangerous Substances

This headline may sound dramatic, but it does not have to be. Obviously, if you are investigating an unknown or potentially dangerous virus, it is highly important that your lab has the necessary clearance to work with this type of virus. In many other cases the potential hazard of a substance may not be quite as obvious. Many research projects may include elements that require extra attention. It is quite common that you find yourself working with potentially carcinogenic or otherwise dangerous chemicals. Working with such things sometimes requires laboratory specific certification and maybe even specific certification for the individual researcher, along the lines of the animal experimental certification as outlined below. Always keep this in mind when designing your study, consult your supervisor to make sure that you hold the necessary certifications and licences needed to conduct your study. You will be surprised how often you will find relatively specific legislation on matters that you did not expect.

9.4 Animal Experimental Certification

In general working with experimental animals requires two things. As a researcher you first of all need to be certified to handle experimental animals. You obtain such a certification by attending courses in animal experimental research. Such courses are available in almost every country and they are more or less identical across the nations. Usually they are available in different levels because individual levels of clearance are required for different purposes. For instance, you often need an initial course to be allowed to participate in handling of experimental animals and conduct experiments together with more experienced researchers. The next level of certification will license you to be the senior researcher, responsible for the safety and ethical aspects, and tasked with gaining all relevant approvals for the project. The courses mainly cover the legislation specific to your country, along with animal ethics and specific animal handling procedures. In most countries there are animal experimental committees that are required to approve your protocol prior to initiation of your project. Bear in mind that such approvals are not handled overnight. The approval process is time consuming, so you should get the process rolling as soon as possible. The animal experimentation licence does not apply to a specific person, but to a specific project. It may, therefore, not be necessary for you to apply yourself. It may very well be a task for your supervisor. But remember that it is very important that you become certified in handling experimental animals yourself prior to starting the actual experiments.

Source: Shutterstock

9.5   Patient Related Research

When conducting research on human beings there are many rules to consider. This is because most data from humans are confidential. Sometimes the amount of legislation associated with clinical trials, and especially drug trials, can be extremely overwhelming. In these cases many universities have dedicated so-called GCP-units that aid in the study design of clinical research. GCP is short for Good Clinical Practice. Their job is to know the details of the local legislation that apply to your study. It is generally advisable, and sometime actually a requirement, that you consult such expertise early in the planning phase of your study to avoid unnecessary delays. Specifically, you should pay attention to rules on removing personal identification numbers, social security numbers or the like before storing the information. There may also be regulations on how to store your data even though personal identifiers have been removed. Bear in mind that you generally must apply for permission to store personal data. Many countries have public data protection agencies that need to be consulted prior to initiation of the study.

9.6   Researching Drugs

Testing drugs, especially in humans, is probably one of the most complex research areas from a legislation point of view. If you are considering a drug trial you need time. You cannot start early enough, formulating your protocol and applying for the needed approvals is a long process. In almost every country you will find a Healthcare Products Regulatory Agency or the like, taking care of approval of all drug trials. It can be a cumbersome task to reach final approval from such agencies, but bear in mind that this is for good reason. You do not want to expose your study subjects to any unnecessary risks.

List of authorities to have in mind:

- Committee on health research ethics, local or national.
- Animal Experiments Inspectorate
- Data Protection Agency
- Good Clinical Practice
- Medicines and Healthcare Products Regulatory Agency

9.7   Collaborations with the Industry

Collaboration with industry is certainly not a beginner's task. You should always have your supervisor oversee such arrangements, but an overall under-standing of the collaboration is still pertinent.

An interesting aspect in the process of taking new drugs or devices from idea to practical implementation is the industrial aspect. You seldom bring a product to clinical use without the involvement of commercial partners who can manufacture and market the product. These companies normally also secure the approval from national and international public certifying bodies like the FDA (The Federal Drug Administration) in USA or CE (Conformité Européenne) in Europe.

The commercial-scientific interaction can enter at any point in the "Global View of Research" (Figure 2.1). The industrial partner can enter the process at any time to play a role in production and marketing. Often researchers and companies work jointly for some part of the journey. Such collaborations should be handled appropriately from both an ethical and business prospec-tive. There are some basic codes of conduct you must acknowledge.

9.8   The Scientific Code of Conduct

As a public scientist, working in a public university or hospital, you should never perform research for a private company where only predefined positive outcomes will be accepted. As a researcher, you do not work to prove that a hypothesis **IS** right or wrong. You must have an open mind and investigate **WHETHER** the hypothesis can be verified or falsified. In scientific terms, both outcomes are valid and interesting.

You must never accept to hide or mask results, which may discredit a product a company wants to commercialise.

Apart from actual salary, you should not receive gifts or donations for your personal or private use for conducting research for a company. As this can potentially lead to subjective and biased interpretation of results.

In order to ensure objectivity you should always collaborate with companies based on a collaboration contract, where the above items have been addressed and discussed to ensure mutual understanding. When these precautions are taken, collaboration between researcher and company can be very fruitful and enriching for both parties.

9.8.1 Commercialising Your Research

Even as a young researcher, you are very much in a position to invent potential future products, which can be commercialised. In such cases, it is pertinent that you ensure a patent application has been submitted for the invention before the scientific results are published and publicly accessible. When you have published your results, it is common knowledge and patent cannot be applied for. Therefore, consider a potential commercial path before you publish your research. You may not want to commercialise your discovery yourself, in that case let others with those preferences take the invention to market, and let future patients benefit from a new product.

On the other hand, you may want to pursue commercialisation of your invention yourself. This is often possible and even encouraged by most universities and public hospitals. In this field, there are also codes of conduct and some approaches are more appropriate than others. You may even go all the way to establishing your own spin-off company in order to manufacture and market new products. This path is complex and in no way a beginner's task—collaborate with more experienced researchers who have taken the commercial path before. Commercialising research is an interesting field of work, but it is a task that only a few researchers will ever face. We will, therefore, not go further into this matter in this book.

10

Applying for Funding

Contributing author:
John Westensee
Deputy University Director
AU Research Support and External Relations
Aarhus University, Denmark

In Chapter 4, we discussed how to define your project, and you learnt that you need resources like office space, equipment, travel money, lab supplies, etc. in order to carry out your project. Your supervisor may well provide some of these resources, but often you will have to find funding from sources outside of your institution to finance elements of your research. This goes for all researchers irrespective of their position on the research ladder. With the drastic increase in the number of researchers both nationally and internationally, the competition for funding is getting more and more fierce. Learning how to construct a strong funding application is, therefore, crucial for any budding researcher.

On average, the rate of success for a funding application is in the range of 5 to 15%. At the same time, around 75% of all proposals are deemed to be of "good" or "very good" quality. This means that around 75% of all proposals are worthy of funding from a quality point of view. These statistics depict clearly the competitive nature of the funding process.

After reading the above passage, you may be left intimidated by the prospect of acquiring funding, but do not be discouraged! This chapter will show you how to prepare a high quality funding application and will provide you with useful tips to give you that all important edge over your competitors.

© Springer International Publishing AG 2017
P. Agger et al., *A Practical Guide to Biomedical Research*,
DOI 10.1007/978-3-319-63582-8_10

Remember you are not alone, in order to be competitive, your application has to be a team effort. So prepare your applications in collaboration with your supervisor or other experienced members of your research group. Their knowledge can be a tremendous help, use it!

10.1 How to Find a Potential Funder

The first challenge is to find out where you can find funding for your research. There are a number of commercial services available, which highlight the funding opportunities available in specific countries. Two examples are ResearchProfessional in the UK, and the SPIN database at InfoEd in the USA.

A good starting point, however, is the Research Support Office at your home institution. They often have access to the above-mentioned resources and will have lists of relevant funding opportunities for you. They generally have detailed insight into which funders you should contact, and will be able to provide guidance on the application process.

Before you start writing your proposal, you need to consider three things:

1. What is the overall agenda of the funder? The goal of many foundations is to solve a problem, and this may be broad in nature or extremely specific. Maybe their mission is to improve the quality of life for a particular patient group, or to specifically help researchers acquire new equipment. This should influence how you write and communicate your application. Also, does the funder actually provide the amount of money you are asking for? Do not ask for money they simply cannot provide.
2. Who will review your proposal? If researchers in your field are to conduct the review, you can write in a more scientific or technical language, and assume a certain level of prior knowledge. But quite often, the reviewers will be lay people. In other words everyday people with no formal scientific training, this may be representatives of the funder, or even the patients themselves. In this case, you should tone down your use of scientific language, and avoid technical jargon, which will alienate the reader. Assume no prior knowledge, and make it as clear as possible.
3. Who should apply, you or your supervisor? As a researcher you are often faced with the question of who should actually make the application. You of course want to give the application the best possible chance of success, but there is no clear-cut answer, and it is often grant specific. Often larger amounts of money fall into the hands of experienced researchers; this is because foundations can be more confident that they will get their money's

worth! Hence, if you are a young researcher, you would normally let your supervisor be the main applicant on larger applications. On the other hand, some foundations may prefer to fund young researchers, but this is often reserved for smaller grants. These grants are usually for activities such as travel costs, specific equipment or courses, and other direct costs. Please do not underestimate the value of small foundations. Approval from one of those shifts your position from "not-previous supported researcher" to a funded researcher. This position improves your chances of raising further funding in the future. "Funding brings funding".

Even though your supervisor may have to be the main applicant you can still participate in the formation of a grant application and we encourage you to do so! There is much to learn, and being a named individual on a grant is great for your CV!

10.2 Application Types

In general, there are two types of applications; the ones with an application form and those without.

10.2.1 Applications with Application Forms

Normally, public funders and large private foundations use a fixed format for proposals. In this case, the foundation controls the application structure and provides specific forms to be completed. You do not, therefore, have the flexibility to compose your own application style, but on the other hand, it ensures that you provide the foundation with the exact information they need. Generally the application consists of multiple sections as depicted in Figure 10.1.

10.2.2 Applications Without Application Forms

Often, small private foundations neither use application forms nor do they have specific proposal guidelines. In such cases, it is a good idea to contact the foundation directly to enquire what they would like to be included in the proposal. Maybe they can provide you with some tips and tricks on how to compose your application. Sometimes you will, however, not be able to get

1. Summary
2. Project description. This is a modified version of your protocol
3. Lay man's description
4. CV
5. List of publications
6. Budget
7. Letter of recommendation from supervisor

Fig. 10.1   The contents of a funding application

such feedback. In this case, you should structure your application as suggested in Figure 10.1. Normally, the smaller private foundations do not need a long and detailed project description. If your application is limited to only a few pages, do not spend too much time outlining the background and methods in the project description. The summary and lay man's description are, in fact, the most important sections.

When writing applications without forms, consider adding a **cover letter**, preferably on official letter paper from your institution. Such a letter should constitute the first page of your application. It should state the objective of the project in clear and simple language and describe what you expect to deliver. It should obviously correspond with the objectives of the funder. You should, furthermore, state how much money you wish to apply for and for what specific purpose. Also state if you expect funding from other funders. Your cover letter should not exceed one page.

10.3 The Contents of an Application

In the following section we will walk you through the specific contents of Figure 10.1. When writing your application, always keep the identity of your target audience in mind: Who will be evaluating your proposal? Will it be lay persons or experts? What level of prior knowledge can be assumed? What do they need to know in order to understand and evaluate your project? And finally, how can you convince them that your project should be funded? Try to find information regarding the evaluation criteria, and adapt your application accordingly. And last, but certainly not least: Always follow the application guidelines of the funder. ALWAYS!

Always follow the application guidelines of the funder. ALWAYS!

10.3.1 Summary

The project summary should be a concise overview covering the most important points in your project description.

The summary should normally be around 200–300 words and answer the following questions:

- What is the problem?
- What do you intend to do about it?
- Why is it important?
- What is the expected outcome?
- How will the work be accomplished?

Do **not** include references, budget information or personal information in the summary.

When explaining the importance of your research, try to strengthen your argument using facts and figures from epidemiology studies, clinical trials and publicly accepted sources like political documents from the EU, UN and OECD. To ensure the highest impact, keep such statements short and simple, for example:

> According to the WHO, by 2050 75% of the population in the developed world will be obese, and cardio-vascular diseases will account for 50% of all health costs. This project will help decrease the societal burden of obesity.

Typically, the reviewers will read the summary to get an overall idea of the project. If they are not convinced by your project after reading the summary, it is highly unlikely they will read any further, and your chance of funding has vanished at page 1. It is important, therefore, to write an engaging summary, which gets the reviewers on your side before they even start reading the main application. This is your chance to "sell" the project. You only get one chance to make a first impression, better make it a good one!

10.3.2   The Project Description

There are quite a few things to address in your project description. The project description is a version of the protocol (see Chapter 8) targeted at the funder. Your protocol is, therefore, an important point of origin when preparing the project description. Especially with small private foundations, the project description is a much shorter version of the protocol, since the funders do not need or want all the details contained in the protocol. The larger the application, the more overlap there will be with your protocol.

In the following section, we presume that you have already produced your protocol (see Chapter 8). We will hence only comment on how you should consider changing and adapting your protocol when applying for funding.

10.3.3   Importance of the Title

The title for the project description does not necessarily need to be the same as the title of your protocol. Sometimes you may want to twist the title a bit to meet the objective of the foundation, and to ensure attention from the foundation board members. The art of writing a good title for your application is very much about selling yourself and your project to raise interest and fascination—with a certain degree of moderation of course!

10.3.4   Keep Introduction and Methods Sections Short and Inspiring

The introduction section may benefit from being shortened a bit. It should focus on putting your project into context, but do not describe your entire field of research. You should mainly mention what knowledge is missing from your field, and demonstrate how your project fills these gaps. Make sure the following issues are covered:

- Outline the current state of knowledge in your field.
- Demonstrate the scientific rationale for carrying out the project.
- Draw attention to your previous contributions to the field.
- Make clear and coherent connections to your project objectives and its potential impact.

Likewise, the "Materials and Methods" section in your protocol will often be too complex for a funding application. Make sure that you have clearly described: How you will carry out the project and demonstrate that you, and your co-workers, have competencies to master your chosen methods.

10.3.5 Lay Man's Description

The lay man's description is a simple, easy-to-understand description of your project. It should be understandable to everybody—even your family and friends! You can use the four questions in Section 10.3.1 above as an inspiration on what to cover in this paragraph. But bear in mind that this document is very different from the summary. The summary is often constructed in the classical scientific way, gradually narrowing down towards the central point. The lay man's description, conversely, should use a more direct journalistic approach. Shoot first, explain afterwards! This is a very effective and forceful writing style that encourages the reader to continue reading. Think of how a newspaper article is structured, with a headline that delivers the entire message, which is then briefly explained in the opening paragraph, often emphasised in bold or italic, and then further elaborated in the main body of the text. It is easy to forget which terms are actually lay men's terms and which are not. It can, therefore, be a good idea to ask a friend or a family member, who has no knowledge of biomedical research, to read through and comment on your lay man's description. Ask for a short summary of your study and you will soon find out if they have understood your message.

Often foundations use the lay man's description as a tool to inform the first round of selection, even before reading the summary. Typically, foundations receive hundreds of applications, and they need to decide which projects fit into the scope of the foundation before doing a full evaluation. The lay man's description can very well be considered the most important part of your application. Make sure it reads well.

10.3.6 The Research CV Versus Other CV Types

The Curriculum Vitae (CV) is a description of all your credentials and qualifications. It serves to convince the evaluator that you have the right academic and managerial competences to carry out the project. Your CV may also help you to decide whether you or your supervisor should be the formal applicant of the project.

As a new—or even not yet started—researcher, your CV will by nature not be very extensive. Do not worry, the funders know what to expect from the CV of a new researcher. Honesty is the best policy, resist the temptation to inflate even remotely relevant past activities. Mention what is relevant, and just put a few relevant headlines for the "not-yet-achieved accomplishments", to indicate your future ambitions.

If your CV is a little light consider including a description of the group you are working in. Include who are you working with and their level of experience. On some occasions it may make sense to include the CV of some of your collaborators.

10.3.7 Composing Your CV

In general, you should use bullet points and present data in reverse chronological order. Prioritise information directly relevant to the project.

An experienced researcher will be able to provide relevant information for all of the headlines mentioned below. These are mentioned here merely to inspire you, but if you can provide any experiences relevant to any of the headings by all means do so! Even initial research exercises such as the writing of your protocol, carrying out pilot studies, or initial lab training, is of relevance. Even though you are composing a research CV you can, in certain situations, benefit from adding non-research related activities as well. If, for instance, you are applying for a research post it can make good sense to include activities that show that you can work independently, have leadership skills or maybe you have administrative experiences? These are competences that are highly sought after in research. You should, however, gradually replace such CV entries with strictly research related competences as soon as you achieve them.

A competent research CV typically includes the following:

- Personal data (name, age, contact information, etc.).
- Academic training and degrees (academic degrees with month and year of completion).
- Professional employment history.
- Major career breaks (paternity/maternity leave, military service, etc.).
- Other scientific qualifications (like presentations at conferences).
- Awards and honours.
- Organisational and administrative duties (project management, leadership in projects).

(continued)

- Research competences/research areas.
- International relations.
- Patents.
- Current research grants.
- Education and supervision of students and postdocs.

In addition to the information outlined above, you may also consider adding your interactions with society, such as non-scientific articles, presentations to the public, and engagement with industry. Note, the order of the headings is not restricted, feel free to reorder them based on personal preference.

10.3.8 List of Publications

As a new researcher you will only be able to provide an empty space where your list of publications is supposed to be found. Again, this is fair enough—do not include essay homework from high school in an attempt to impress. When your research career matures, you will include publications as some of the most important and meriting achievements. Every famous researcher started with his or her first publication—before that, they had none!

You should only add publications to your list, which have actually been published or accepted for publication. As your career progresses you should aim at including the following in your list of publications:

- Peer-reviewed publications such as articles, monographs, proceedings with a referee and book chapters.
- Non-peer-reviewed publications.
- Patent references.
- Other types of publications.

Add data like authors, year, title, place of publication, volume number and first and last page number, or article number including number of pages. You should, furthermore, add the DOI (Digital Object Identifier) and pubmed ID so the paper is easy to find. Finally you should state the number of citations your paper has received so far. Mark your name clearly with underlining or in bold. See Figure 10.2.

P. **Agger**, RS. Stephenson, et. al. "Insights from echocardiography, magnetic resonance imaging, and microcomputed tomography relative to the mid-myocardial left ventricular echogenic zone." *Echocardiography. 2016 Oct;33(10):1546-56. doi: 10.1111/echo. 13324. PMID: 27783876. Citations: 1*

Fig. 10.2 An example of how to enter a publication into your list of publications. The author list can be shortened with "et al." after your name has appeared. Highlight your own name. Include the title of the publication and the full reference to the journal including DOI, Pubmed ID and the number of citations

10.3.9 Budget

In Chapter 8 "The Scientific Protocol", we outlined how to incorporate a budget into the protocol itself. But since the budget is the key message when applying for funding, it warrants its own document. As in the protocol, your budget should be as realistic as possible. You are normally required to include all cost relevant to your project, even though you may only be applying for part of the budget in the application. This shows that you have thought about all aspects of your project. In particular, it is relevant to include costings such as conferences and publication expenses. The foundation will often ask you to provide reports on your expenses at specific points throughout your project, so it is important to keep track of your finances. Often the department secretary or financial team will aid you in this process.

Take your short budget from your protocol and elaborate thoroughly. Remember to include overheads in this budget if the foundation permits you to do so. Overhead is a term for the indirect costs connected with completing your project. Overheads are calculated as a percentage of the project's direct costs. The percentage varies from foundation to foundation and can also depend on the institution. Always use the percentage indicated in the foundation's call for proposals. Bear in mind that many private foundations specifically refrain from payment of institutional overheads.

Often the foundations ask you to elaborate on grants that you have already received for your study. It may seem counterintuitive to state that you have already been given money, when you are in fact applying for more. One may think that the foundation will refrain from supporting your research if they know that you have already been given your share. Nothing could be further from the truth! In fact, it is the other way around. If a foundation sees that other foundations have considered your study worthy of support, they will be more inclined to support you as well. If you have already been granted a large portion of your budget it is often very attractive for another foundation to provide you with the rest of what you need. In that way, they can be confident you will complete your study and publish your work with their name in the acknowledgement section, and this type of exposure is exactly what they like to pay for.

10.3.10 Letter of Recommendation

Especially as a young researcher, you should always include a letter of recommendation from your main supervisor. This document could be what convinces the foundation that their money is well spent in your hands. The letter should indicate that your supervisor fully supports you and will provide the help you need to complete the project. Furthermore, it should state why you are the right person to conduct the study and that your supervisor also considers your project highly relevant. Of equal importance is a good description of your planned role in the study. If the foundation allows it, include multiple letters from several collaborators.

10.4 A Couple of Final Tips and Tricks

When writing an application, you need to keep the funder in mind. What are the needs and interests of the funder? You are marketing your idea and, therefore, targeted communication is essential. Keep your application short and simple. It is not a competition to show how clever you are. You need to think of your target group—evaluators are human beings, and researchers serving as evaluators are often required to read many proposals. Help them make the right decision quickly.

Finally, we provide a list of 10 common causes for rejection of funding applications. Hang this list on your wall as a reminder when you write your application.

- Poor match between research idea and focus of the funder.
- Weak summary.
- Goals and objectives are unclear, over ambitious, unachievable (feasibility).
- Weak argumentation for relevance of the study.
- Unclear project description and work plan.
- Poor organisation of text.
- Long phrases, jargon, abbreviations.
- Deviations from foundation guidelines.
- Ignoring review criteria.
- Writing solo without help from supervisor or colleagues.

11

Data Handling

The main objective of research is to gather data that can answer your research question. But what is data and how should you handle it after you have collected it? There are both practical and legal issues to be considered. The present chapter provides guidance for effective data structuring and handling, to ensure that you do not lose your valuable data or let it fall into the wrong hands!

11.1 What Is Data?

When we refer to the term data it covers a very broad palette of information. It constitutes every little piece of information you gather with the purpose of answering your research question. It can obviously consist of text files or spread sheets, but it could also be photos of experimental set-ups or clinical imaging data such as X-rays or scan data. Data may come in digital form or in hard copies. Maybe you have asked a group of study subjects to complete a questionnaire, this could be online or as a physical form. Collecting data is often very straight forward, but it is important to consider the specific handling of the data, both short-term and long-term, before you start.

© Springer International Publishing AG 2017
P. Agger et al., *A Practical Guide to Biomedical Research*,
DOI 10.1007/978-3-319-63582-8_11

11.2 Structuring Your Data

It is extremely important to put some thought into how you structure your data when you store it. In the beginning of the project life cycle it may seem easy to maintain an overview of your data, but during the course of our study you are likely to struggle with maintaining that overview. You need to have an appropriate structure. Start considering data structure from the start of your project. Make sure that your data is organised logically, not only for yourself, but also for others. In the unlikely event of you not being able to complete the study, another researcher may have to take over the reigns; this is impossible if they cannot navigate your data structure. For advice on how to design your data structure it is very much advisable to consult some of your younger colleagues, the senior supervisor has probably not dealt with raw data for a long time. Every bit of data must be easily found and the nature of the data must be crystal clear. A common mistake is to create several versions of your main data set as you progress with your data analyses. This makes it impossible for outsiders to know which data are actually the most up to date. Preferably maintain a single data file that is named appropriately. Do not label datasets with suffixes such as "old", "new"; you always end up mixing something up. If you have to name data files by date use the format YYYY-MM-DD with the year preceding month and day. In this way your files will be ordered chronologically when you ask your computer to sort them by name.

It is difficult to find your way around when many files are stored in the same folder. Using a tree of subfolders with carefully considered names can help you overcome this problem (Figure 11.1). Keep your data in a folder structure that is easy to follow, so that both you and your collaborators are capable of navigating the data. There are many solutions to this challenge. Our suggestion is to structure your folder tree in the same way as your scientific manuscript. The IMRAD structure can help you in this situation as well. IMRAD is an acronym encompassing **I**ntroduction, **M**ethods, **R**esults **A**nd **D**iscussion. This is a commonly used way of structuring scientific communications. You will come across it several times in the remainder of this book. Figure 11.1 shows how this concept can be applied to create a folder structure that is clear and easy to work with.

Put data that relates to the introduction such as the protocol and the references in the introduction folder, data on courses you plan to attend and documentation for your experimental setup in the materials and methods folder. Your actual study data obviously belong in the results folder, as does any representations of the data that you make during your data analyses such as plots and tables. In the discussion folder you should place everything

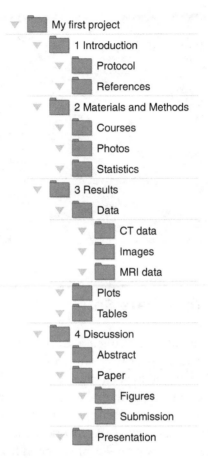

My first project
   1 Introduction
      Protocol
      References
   2 Materials and Methods
      Courses
      Photos
      Statistics
   3 Results
      Data
         CT data
         Images
         MRI data
      Plots
      Tables
   4 Discussion
      Abstract
      Paper
         Figures
         Submission
      Presentation

Fig. 11.1 An example of how you can build your folder structure on your computer by the use of the IMRAD structure

that concerns dissemination of your work, such as papers in preparation and presentations for meetings or conferences. Be generous with the number of subfolders. Subfolders often make it easier to navigate your way through your data and quickly find what you are looking for. You may find it advantageous to keep several copies of your files in several different folders for different purposes, but this is not in any way ideal. Keep only one version of your files and keep a file system that you can remember and maintain effortlessly. Obviously you should always maintain a backup of your files. This is not the same as keeping multiple copies of your files in the same folder on the same computer. In the following we will elaborate on the concepts of keeping a good backup of your data.

Source: Shutterstock

11.3 Keeping Your Data Safe

Accidents happen! Computers get stolen and they might occasionally break down. Naturally it always happens to other people and not to you, right? But guess what, it might just as well happen to you. You should, therefore, prepare for the worst, and start out from the very beginning with a good plan for keeping a backup of your data. In science the consequences of losing your data might mean losing months or even years of work, some of which can never be reproduced. There are several ways of keeping a good backup. Either solution may be as good as the other, but at least it should adhere to this general rule: always keep at least two backups; one local and one remote. The simplest and probably cheapest solution is keeping a **local backup** as an external hard drive, to which you copy your data at the end of a working day, while the **remote backup** can be a similar hard drive at home. The idea is that you always have a backup that is not located beside your computer. Then the risk of someone stealing your computer and both hard drives simultaneously is next to nothing. This solution, however, may be quite laborious and you may very well find yourself forgetting to backup your data.

The ideal backup solution should, therefore, be fully automated. There are several Internet based solutions that perform automated continuous backups to remote servers. This is probably the most ideal backup solution for research purposes. There are hundreds of server based providers and it would not make sense to describe any of them here. Some companies provide combined file sharing and backup solutions, some provide unlimited storage space, and others focus on low pricing and even free solutions. The opportunities are numerous. The easiest way of getting an overview is to search the Internet for "Online backup review". Most companies offer a free trial so that you can try out before you buy.

What if you obtain your data in hard copy for instance forms and questionnaires or print outs from laboratory equipment? You should never store your data only in hard copy. Preferably collect your data digitally whenever possible or at least digitise it as soon as you can. Scan printouts or type them directly into a spreadsheet. Be aware though that there may be laws in your country preventing you from using public backup services if you intent to store confidential information that can be traced back to a specific person. Consult your supervisor on local legislation before deciding on a backup solution. When all of the above is settled you can rest assure that your data is safe and the only trouble you will encounter from a computer breakdown is the trouble of fixing your computer or buying a new one.

12

Data Analysis

Ok, so you have finished collecting your data; now it is time to find out exactly what your data shows. You finally have the opportunity to answer that research question you posed all those weeks, months or even years ago! We do this using data analysis.

Regardless of whether it is a preliminary pilot study or a fully completed research project, it is important you approach the analysis of your data in the same way. Do not waste your time using unsuitable and invalid analysis methods, you will only end up having to redo the analysis, or worse still, you will produce false results and make misinterpretations. Imagine claiming to have made a significant contribution to your field, just to have your results invalidated by your peers based on a poorly chosen analysis method.

You should look upon analysis of your data as an intriguing prospect, you are about to unlock the true meaning of your results, and who knows, you may reveal some unexpected treasures along the way. That novel finding could be just around the corner! Before you know it you could be contributing truly new information to the scientific literature. In this chapter we will discuss the importance of appropriate tools in data analysis, provide basic advice, using specific examples, on where to start, and finally you will receive insight into how to interpret the results of your data analysis.

© Springer International Publishing AG 2017
P. Agger et al., *A Practical Guide to Biomedical Research*,
DOI 10.1007/978-3-319-63582-8_12

12.1 Work Flow for Data Analysis

Having a structured approach to your analysis is essential, the flow chart in Figure 12.1 depicts a classic systematic approach, which can be applied to almost any type of data set. You should always start by collecting your data in one place—*Collate*. Next, carry out basic descriptive analysis to get an overview of your data—*Describe*. Then present the initial results graphically to get an overall idea of the nature of your data—*Depict*. Now for the moment of truth, how can your analysis be interpreted? Is the observed difference statistically significant?—*Compare*. Then finally, a comprehensive illustration of your research message—*Visualise*. Use this flow chart to guide your reading of this chapter and when planning your analysis protocol.

12.2 Deciding on Appropriate Tools

Data analysis can be performed in countless right ways, and countless wrong ways. Just because you get the result you were looking for, does not mean the method you are using is valid! The majority of analysis methods will produce results that appear correct regardless of whether it is a valid test for your specific data. You must, therefore, ask yourself, is my chosen method fit for purpose? And does it answer my research question in a valid way? Focus, therefore, on

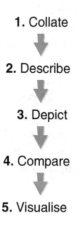

1. Collate

2. Describe

3. Depict

4. Compare

5. Visualise

Fig. 12.1 Work flow for data analysis—(1) Collate—Collect your data in one place. (2) Describe—Perform basic descriptive analysis. (3) Depict—Present the initial results graphically. (4) Compare—Conduct and interpret comparative analyses. (5) Visualise—Final illustration of your research message

finding the right method, not the one, which gives you the results you want to see. Often the simple option is the most suitable, try not to get lost in the plethora of possibilities. One way to avoid this is to consult with a colleague or statistician.

When searching for the right analysis method, note you are not always required to produce an entirely new methodology. You have a wealth of information at your fingertips, use the advice we provide in the Chapter 5 to tap into the scientific literature. Remember, any original scientific manuscript, brief communication or technical note will provide a detailed description of their methodologies, including their data analysis. With further in-depth information often found in the data supplement. It may well be the case that nobody has conducted your exact study previously but this does not mean you cannot look in the scientific literature for inspiration. Chances are the type of analysis you need is already published. This is a huge advantage, not only does it provide you with your method, but because it is published it has already been through the peer-review process, and has hence been validated by your peers. Another option is to look a little closer to home, often your lab or colleagues will have analysed similar data in the past, and chances are they will already have analysis methods in place, do not hesitate to ask around.

It is worth noting that you should always strive to have a good understanding of your chosen analysis methods, and be able to describe why you chose them, and what they measure. Questions regarding the suitability and validity of your analysis methods are always popular following any poster or oral presentation, be prepared to defend them!

Source: Shutterstock

12.3 Spread Sheets, Graphics and Statistics

Data are most commonly described as either qualitative or quantitative, but what does that mean? Simply put qualitative data is non-numeric, it is often visual in nature, examples in the biomedical field include histology, medical image data, and patient photographs, but it can also constitute diary accounts, or answers to open-ended questionnaires. Although they can be quantified using post-processing, inferences based on interpretations of the raw data are inherently non-numeric, descriptive and subjective. Quantitative data on the other hand is numeric, it can be categorised, ranked and statistically analysed. The data is often presented in tables and graphs, which display your numerical message.

12.3.1 Steps 1 and 2—Collate and Describe

Data analysis is very much data specific, and often specialised. There are, however, some initial strategies, which can be of use to get a general overview of your data. Whether you have conducted quantitative analysis of your qualitative data, or you already have quantitative data, you should start by collating your data in one place (step 1—Figure 12.1). Spreadsheets are the best place for this as you can easily categorise, rank and arrange your data in a cellular format. Most types of spreadsheet software allow you to use self-written formulas and have integrated statistical packages. Start simple, often very basic analysis can be extremely informative. Calculate the mean, mode or median and investigate the accuracy of your data by calculating the standard deviation. Standard deviation measures how much your data deviates from the mean and is used in the final written and visual depiction of your data (step 5—Figure 12.1). Applying these simple analysis methods will provide an initial overview of your data, which will allow you to get a feel for what your data is showing, and help you plan more in-depth analysis methods. These simple approaches constitute step 2 of your work flow and can be defined as descriptive analysis methods (Figure 12.1).

12.3.2 Step 3—Depiction

You can next consider graphical representation of your data. Plotting the distribution is always a good idea. Finding out whether your data is normally distributed or not is important, since it dictates the types of statistical analyses you can use. Normally distributed data can be analysed using the so-called parametric methods, which are considered more powerful. Distribution is often presented as a histogram, whereby your data is categorised into self-defined intervals, and the frequency of values in each interval is represented by the height of the designated bar (Figure 12.2). For example, you may use this method to interrogate the distribution of age within a patient cohort. Inherently the data is centred around the mean. In the case of normally distributed data (Figure 12.2a), a fitted curve has the classic bell-shaped appearance. Conversely in non-normally distributed data (Figure 12.2b), the curve will be skewed.

Scatter plots are a great way to get an overview of how coherent your data is, in other words how closely your data adhere to a linear trend line (Figure 12.3). It also allows you to easily identify anomalies or outliers (red circle, Figure 12.3a). In this case you will plot two variables against one another, for example, the age of a cohort of patients against the frequency of visits

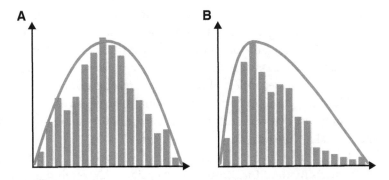

Fig. 12.2 Illustrative histograms showing normally distributed (**a**) and non-normally distributed (**b**) data sets

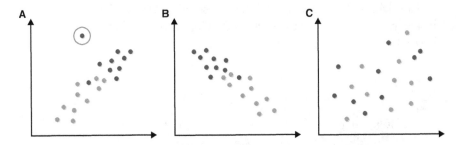

Fig. 12.3 Illustrative scatter plots showing two coherent data sets (light blue and dark blue) with positive (**a**) and negative (**b**) linear relationships, and non-coherent data with no relationship (**c**)

to the emergency department. Figure 12.3 shows scatter plots from data sets with a positive linear relationship (a), a negative linear relationship (b) and no relationship (c). Scatter plots also allow you to appreciate differences between studied groups, which can guide future in-depth analysis. For example, if you wish to plot your two variables, and investigate the difference between males and females. Study Figure 12.3 once more. There may well be a significant difference between the two study groups (light blue and dark blue) in panels (a) and (b), but is it highly unlikely further analysis would find a significant difference between the data presented in panel (c). Many types of software allow automatic correlation analysis of scatterplots in which the coherence and relationship between variables can be assessed and quantified. We suggest you refer to relevant literature if you wish to learn more about such methodologies.

12.3.3   Step 4—Comparison

There are many types of statistical software out there to help with step 4 of the analysis work flow (Figure 12.1), as a general rule you should use the one you understand and can navigate the best. Many statistical tests are standardised and thus will produce the same output regardless of the software. For example, a Student's t-test used to assess the difference between two variables will generate the same result regardless of the software in question. But again, you should consider using the same software as your co-workers, this means any required training can be provided in-house; furthermore, it aids inter-group handling and analysis of the data.

So what factors do you need to consider when it comes to deciding on the correct analysis method for comparing your data? Although they can help with the process, many of your colleagues, including the most experienced ones, are in fact not qualified to decide on the most valid statistical test for your data. Many departments will have designated statisticians, use them! It is always advisable to ask yourself the following questions before seeking help, and especially prior to visiting your resident statistician.

- First, was my data collected using valid methodologies?
- Should you expect variation or anomalies?
- What is my sample size?
- How is my data distributed?
- How many variables do I want to test?
- What confidence levels do I wish to set?

They will appreciate your preparation, it shows you are willing to contribute to the analysis and actually have an interest in the process, and you are not just looking for someone to do all the work!

We have mentioned previously the importance of a multidisciplinary approach to research. Comparisons can also be made between your quantitative statistical analysis and qualitative data. For example, you may wish to compare the statistical analysis of the time from an initial leg fracture to recommencement of load bearing, with a series of X-rays over time from the same patient cohort. This type of multidisciplinary comparison, in which different types of data are compared, is inherently subjective, but can help cement your message, and can be considered as subjective validation of your statistical findings.

12.3.3.1 Verify or Disprove Your Hypothesis

Is the comparison statistically significant? You may have seen many manuscripts refer to P-values, you may have even encountered the term when questioned about the statistical significance of your work. But what is this P-value everyone is talking about?

P-value is short for probability value, it is a product of most statistical tests, and they allow you to directly and subjectively verify or disprove your hypotheses. Specifically they allow you to either accept or reject your null hypothesis. Your null hypothesis simply attributes any difference observed in your comparisons to chance. P-values range between 0 and 1.0, and simply put a low P-value allows you to reject your null hypothesis, while a high P-value means you have to accept it. For example, if $P = 0.05$, there is a 5% chance that any differences you see between groups is due to chance, you can therefore reject your null hypothesis with 95% confidence. Alternatively, if $P = 0.5$, this tells you there is a 50% chance any differences are due to chance, in this case you must accept your null hypothesis. In other words, if you repeated the experiment there would only have a 50% chance of reproducing the same results.

You will be asked to provide a **confidence level** prior to running your statistical test. Alpha levels relate to confidence levels and can be calculated as your confidence level subtracted from 100%. For example, if you want to have 95% confidence when rejecting your null hypothesis, set your alpha level to 5% or 0.05. In this case if the observed difference between groups is statistically significant your P-value will be $p \leq 0.05$.

An Example

Your study hypothesis is as follows: "liver tissue volume is increased in response to an acute alcohol insult". You have shown your data to be normally distributed by producing a nicely bell-shaped histogram. You proceed to investigate whether there is a statistically significant difference between your two experimental groups. You decide you would like to have 95% confidence when rejecting your null hypothesis. You, therefore, have two possible scenarios $p \leq 0.05$ or $p \geq 0.05$. If your P-value is less than 0.05, you can reject your null hypothesis with 95% confidence, and accept your project hypothesis. If your P-value is larger than 0.05, you must now accept your null hypothesis; "any difference in liver tissue volume between groups in response to an acute alcohol insult is due to chance".

A statistically significant difference can be presented textually as follows:

Liver tissue volume was significantly increased in individuals subjected to chronic alcohol insult compared with controls (10.9 ± 2.36 cm^3 *vs* 7.44 ± 2.28 mm^3) (p = 0.004).

Here 10.9 indicates the mean liver tissue volume in your experimental group, and ±2.36 indicates its standard deviation.

It should be noted that all P-values are important, regardless of whether they fall into or very close to the "significant" interval. Negative results are also informative, but are negative results publishable? The short answer is yes, if they disprove or do not match a previous study this is important data.

But what about borderline P-values, in other words P-values very close to your significance level? In this case you should quote the specific value. There is, however, more to statistics than P-values. They are nice to have, but can also mislead you. Even minute irrelevant differences between groups can be statistically significant if the groups are large enough. Conversely, initially striking differences can be rendered non-significant if the groups are too small. Always provide the descriptive statistics such as means and standard deviations along with your p-values and leave it to your readers to decide the significance of your finding. P-values are generated here in step 4, but are also an important aspect of step 5 (Figure 12.1).

12.3.3.2   Data Interpretation

Correct interpretation of your data is arguably the most crucial aspect of step 4. In biomedical research, it is always important to approach interpretation of your data with a certain degree of common sense and objectivity. Statistical analysis, regardless of the apparent significance, does not tell you anything about the importance or clinical implications of your findings. This is something you determine by looking at all the information your analysis has to offer. Do not just rely on P-values.

Consider, for example, a study where you have compared blood pressures between two groups of patients and you have found a difference between the groups of 3 mmHg, p $=$ 0.02. Given that your P-value is less than 0.05 the difference is without doubt statistically significant, but is it also clinically relevant? It is highly unlikely that a difference of just 3 mmHg has any relevance at all. Maybe your result is a simple coincidence or maybe your sample size is so large that you are able to detect even the smallest difference. In this case, it is important to remain objective.

12.3.4 Step 5—Final Visualisation

A comprehensive illustration of your research message is mightily important, as it will be the means by which peers assess the validity and significance of your findings. Such visualisations will also be the means of conveying the findings of your research in poster and oral presentations, and also in scientific manuscripts. They need to be clear, concise and be able to stand alone.

In step 5 you will build a picture, which encompasses the previous steps of the data analysis work flow (Figure 12.1). Start by presenting the initial difference graphically. Next add information regarding the variability of your data, for example using error bars. Finally, add the P-value, so the reader knows whether the message you are conveying is statistically significant. An example of such a visualisation is given in Figure 12.4. So what exactly does this figure tell us? First of all it is a bar chart, which compares two groups. Secondly, the size of the bars tells us the two groups are different. What about the variance of the data? Well, the different heights of the error bars tells us one group is more variable than the other, but crucially they do not overlap, this means the difference is likely to be statistically significant. The P-value confirms this fact because it is 0.002, and we can therefore be more than 95% confident that this difference is a real phenomenon and not a coincidence.

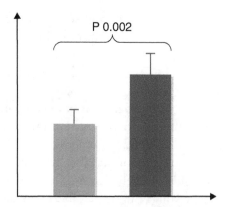

Fig. 12.4 Illustrative bar chart depicting step 5 of the data analysis work flow. The size of the bars indicate the difference between two experimental groups; the error bars indicate the level of variance within an individual group. The P-value indicates the statistical significance of the observed difference between the groups

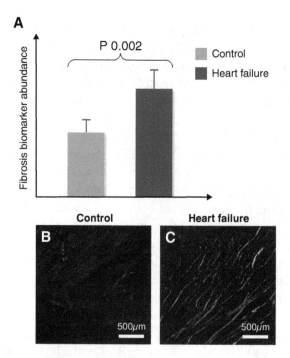

Fig. 12.5 Enhance your research message by supporting your quantitative data with illustrative qualitative data

As described above, if you are able to accompany the message provided by your quantitative and statistical analysis with visual depictions, this can really help cement your message. For example, in a study of fibrosis in the failing heart, numerical statistical analysis of biomarker up-regulation in disease (Figure 12.5a) can be supported with histological depictions of fibrotic distribution in regional biopsies (Figure 12.5b,c). When using qualitative data to enhance your message, ensure the reader knows everything they need to know about the data. Have you included a scale bar, a detailed description of where and how the data was acquired, along with the anatomical orientation of the data? Give your reader the best chance of understanding the relevance of this important accompaniment.

Part III

Presenting and Publishing Your Research

This final part of the book deals with what comes after your actual study. This relates to the various means of publishing your work and using it to further develop your research career through quality networking.

13

The Scientific Conference

As a researcher, you should always strive to be the first in the world to present new research outcomes. It is a great achievement to be the second person to climb Mount Everest, but everyone will always refer to the one who reached the summit first!

The fastest route from conceiving new research results to being the first to publish, is to present your project and your scientific message at a scientific conference.

Scientific conferences vary greatly in size. Some comprise less than a hundred participants, while the larger international conferences can include several thousand participants. Joining a scientific conference can, therefore, be an overwhelming experience. Regardless of its size all conferences are usually structured in the same manner. The same overall concepts apply, and gathering a little knowledge on the basic concepts of conferencing prior to attending will improve your outcome of the conference significantly. This chapter is designed to do exactly that.

13.1 The Conference Abstract

In order to present your work at a conference you need to submit an **abstract**, which needs to be approved or accepted for presentation. Essentially an abstract is your application to present at a conference. It is a brief summary of your project and its results.

The period from submitting your abstract to actually giving your presentation at the conference varies, but is normally 3–6 months. The time

© Springer International Publishing AG 2017
P. Agger et al., *A Practical Guide to Biomedical Research*,
DOI 10.1007/978-3-319-63582-8_13

to publish an article in a scientific journal can vary from 7–8 months to almost 2 years. The process of publishing your work at a conference is brief and uncomplicated, so there is no excuse not to give it a go! There is very limited possibility to present details, neither in the abstract nor during your presentation, but on the other hand you can interact with other fellow researchers working in your field and get valuable feedback on your work through discussions with them.

Presentation of your work at a conference is, therefore, a quick way to disseminate your research, and at the same time, it is a unique opportunity to meet, talk and network with colleagues working in your area in other research groups at an international scale. Especially as a young researcher, attending a conference is a very stimulating activity, where you are introduced to a specialised scientific community. By all means use that opportunity!

There are thousands of scientific conferences worldwide every year. Selecting the right ones is a matter of experience. Your supervisor is critical in that selection process, so ask him or her. Once you have made your choice of conference, it is time to do some research on when and where the conference is held; what are the topics and when is the **Deadline for Submission of Abstracts**. This is an absolute deadline! If you do not submit the abstract on time it will not even be considered for inclusion in the program. Therefore, be sure to be well prepared and make your submission in good time.

On the website for the society organising the conference you will always find guidelines for the abstract submission procedure. An abstract usually adheres more or less to the IMRAD structure as outlined in Chapter 11, but the specific guidelines must always be adhered to very strictly—even down to the smallest detail: The length of the abstract (e.g. 2000 characters **including** spaces), the font type and size, and use of illustrations and tables. There can be even more restrictions and requirements. If you do not comply, you risk your abstract being rejected merely for formatting reasons.

Writing an abstract is an art in itself. You must provide your rationale, results and the key message in extremely few words. Writing a long description is easy; writing the same information in a shorter form is much more difficult. As with anything else, writing an abstract is considerably more difficult the first time than the fiftieth time. To illustrate how an abstract can be composed see Figure 13.1.

Do not underestimate the time it can take to produce a very good and concise abstract, be prepared to work your way through multiple drafts. When you have worked with your abstract for some time and you cannot see how the abstract can be improved or formulated to meet the word limit without losing

MONITORING THE DEVELOPMENT OF RIGHT VENTRICULAR HEART FAILURE USING HYPERPOLARISED 13C MAGNETIC RESONANCE IMAGING.

Peter Agger[1,2], Janus Adler Hyldebrandt[2,3], Esben Søvsø Szocska Hansen[4], Farhad Waziri[5], Camilla Omann[5], Nikolaj Bøgh[5], Christoffer Laustsen[6]

1) Aarhus University Hospital, Dept. Pediatrics, Aarhus - Denmark. 2) Aarhus University Hospital, Dept. of Clinical Medicine, Aarhus - Denmark. 3) Akershus University Hospital, Dept. of Anaesthesia and Intensive Care, Lørenskog – Norway. 4) Aarhus University Hospital, Dept. of Anaesthesia and Intensive Care Aarhus-Denmark. 5) Aarhus University Hospital, MR Research Center Aarhus-Denmark. 6) Aarhus University Hospital, Dept. of Cardiothoracic and Vascular Surgery Aarhus-Denmark.

Background
Right heart failure is common in patients with congenital heart disease, but the evaluation of right ventricular function is complicated and inaccurate and the correct timing of intervention is a matter of continuous dispute. Current diagnostics focus on alterations in anatomical and physiological parameters as predictors of clinical outcome. Magnetic resonance hyperpolarisation, using pyruvate as an active metabolic tracer, is a novel technique that can image the metabolism in tissues non-invasively in real-time. We hypothesised that the metabolic alterations known to occur in heart failure can be assessed and monitored with this technique making it a potential powerful tool in heart failure diagnostics.

Materials and methods
At baseline five female 5 kg piglets were subjected to banding of the pulmonary trunk. Thereby, progressive heart failure was induced by natural growth. At baseline and every four weeks for 16 weeks the animals were assessed with hyperpolarised pyruvate, conventional MR imaging, 4D echocardiography and blood sampling. Finally, at week 16 cardiac physiology was invasively assessed with conductance catheter technique. End-point data was compared to a weight matched control group.

Results
At 16 weeks right ventricular end systolic pressure-volume relationship, dP/dt_max and preload recruitable stroke work were increased ($p<0.05$), while the left ventricle exhibited significant diastolic impairment. The myocardial lactate/bicarbonate ratio, depicting the balance between anaerobic and aerobic metabolism, increased from 0.036 to 0.10 ($p=0.004$). No changes were found in the functional echo-derived right ventricular parameters such as TAPSE, FAC and strain.

Conclusions
At 16 weeks, the animals had developed compensated right heart failure accompanied by a significant impairment of left ventricular diastolic function. Heart failure at this level was not detectable by neither conventional MR imaging nor echocardiography whereas hyperpolarisation data showed an increase in lactate/bicarbonate ratio indicating a shift towards anaerobic metabolism. In this study hyperpolarisation was capable of detecting developing right heart failure prior to conventional imaging modalities.

Fig. 13.1 An example of a conference abstract

information, then you need someone to look at the text with fresh perspective. Now is the time let your co-authors contribute. During the entire writing process, it is pertinent to remember that the evaluators only have very limited time to assess each abstract (maybe 2–3 min). Your abstract must therefore be easy to read, for an expert.

In Chapter 17 we elaborate on scientific writing. We advise you to consult that chapter prior to writing your abstract. Here you can find a few tips and tricks, which apply more specifically to the creation of an abstract.

> • Give your abstract an appealing title, which is easy to grasp and which indicates the message of your abstract.
> • Avoid the use of abbreviations, they slow down speed of reading.
> • Make sure to align your stated study aim and your conclusion. The conclusion shall be a direct answer to your aim. In this way, you ensure that you provide the reader with a clear message in your abstract.

For most conferences, you submit the abstract electronically to the conference secretariat via the conference website. You will find instructions provided here, and it normally functions quite well. At this stage, you can sometimes indicate whether you prefer your presentation to be as an **oral presentation** or as a **poster**. This is an indication of preferences and not a commission. You cannot expect the organisation committee to comply with your preference. Often conferences include far more posters than oral presentations, hence only the best studies are presented orally. Statistically, you are, therefore, far more likely to be granted a poster presentation. The secretariat performs a quality check of the format of your abstract and then allocates it to be evaluated and rated by **external peer reviewers**.

Based on the external expert reviewers' rating of your abstract, the conference organising committee decides whether your abstract is accepted to be included in the program or rejected. The latter fate is a non-appeal decision.

When all accepted abstracts have been allocated to time slots in the program, the conference organisers send information to you about the time for your presentation (see Figure 13.2). The first time you have an abstract accepted to a conference it is something to celebrate!

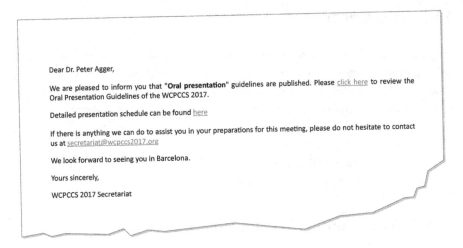

Fig. 13.2 An example of a letter of acceptance of an abstract. Note the links to further instructions on how to prepare

13.2 When Your Abstract has been Accepted

You will now be busy preparing your presentation for the conference, either as a poster or as an oral presentation. Again, it often takes more time than you expect especially the first few times. In this process help and guidance from experienced fellow researchers and your supervisor is critical (see Chapters 14 and 15).

Another part of the preparations is to book travel tickets, hotel accommodation, registration to the conference, etc. All these elements have cost implications. If your supervisor is not in the lucky position to be able to support you with payment of the expenses, you must write applications for funding to cover the expenses (see Chapter 10). Various foundations and institutions support young researchers to attend conferences. Again, the supervisor is the key person for identifying resources to cover the expenses. A major challenge is the often very short time frame from approval of your abstract to the start of the conference. Therefore, prepare applications when you submit your abstract to the conference, so immediately after the abstract is accepted, you can submit funding applications with no delay. It is such a pity to have an abstract accepted to a conference, and not have the money to attend the conference. It is a stressful period, but solutions are normally found. Please also remember that having an abstract accepted for a conference is also prestigious for your research group, department and institute. There are, therefore, always persons around you to ask for help.

13.3 The Conference Setting

During the conference, ensure you make the most of every opportunity. You will be rubbing shoulders with fellow researchers and even the big icons you have read about in the scientific literature, they are all there! You have a unique opportunity to network and establish relations with future collaborators. Most researchers are very open to invite you to visit their research units. If possible, use this opportunity. It broadens your scientific horizon and always spurs inspiration for new collaborative research. Who knows, you might have just met your next supervisor or new employer!

13.3.1   The Scientific Session

Typically a conference is divided into sessions, and often each session holds a specific topic. Your presentation will be assigned to a specific session. Here it is generally a good idea to attend the entire session from the beginning. Knowing what others have discussed prior to your talk is often very convenient. One or more so-called chairmen or moderators are in charge of each session. The main tasks of the moderators are to get through the session within the allotted time frame and to ascertain a fruitful discussion after each presentation. Often the chairperson will provide the first question to get the discussion going. It is always a good idea to research the chairpersons', field of expertise. This could give you a clue on the line of questions they may pose to you, after your presentation. It is the chairman's responsibility to keep time. They are, therefore, allowed to interrupt you and demand that you finish your talk in case you run overtime. But of course it is completely unnecessary to emphasise this because you are going to stick to the time frame—right?!

13.3.2   Dressed for Success

A common question when attending a conference for the first time is: How should I dress? As always you should signal that you are a serious researcher. Showing up in Bermuda shorts or trainers does not convey that message. That said, the dress code can vary a lot between research environments. At some conferences you are frowned upon for not wearing a tie as a man, at others you may be considered too formal. If in doubt go formal. You can always downscale when you arrive and lose the tie or the jacket if you realise you have overdressed. As a woman it can be a little more complex because dressing down "on-site" might not be that easy. Try to plan an outfit that can be downscaled if necessary and bear in mind that in certain countries there may be cultural etiquette to adhere to. Bare shoulders, short skirts and low neck lines may not be generally acceptable. Obviously such measures may make you stand out from the crowd, but not always in a positive way. Go for a dress that signals style and professionalism, and think about which shoes you wear, you will be walking A LOT!

13.3.3   After the Conference

A critical part of establishing and maintaining networks is to nourish the relations. As soon as you come home from a conference, take your time to follow up on the most interesting people you met and with whom you may want to interact later. They will remember you immediately after the conference, but not 6 months later!

Most likely, you will present your first scientific work in a brief form as a poster or an oral presentation at a conference. Next step is to write a scientific manuscript meant to be published in a more comprehensive form in a scientific journal. If you have not already written the manuscript prior to the conference, then you should use the opportunity of being inspired from the conference to finish the manuscript and submit it to a scientific journal. You will find thorough guidelines to writing a scientific manuscript in Chapter 16.

14

The Poster Presentation

Effective presentation and communication of your results is an essential skill in science. Poster presentation is a great way to introduce yourself to the world of scientific communication, and it offers the opportunity to gain feedback on your work from outside your immediate research circles. It also provides vital experience in responding to probing research questions. The presentation style of a poster is the one in which new researchers probably have most difficulties relating to. It likely differs considerably from what you have tried and experienced earlier in your education. You will, therefore, need a bit of help getting started. In this chapter we will present you with some tips and tricks for designing and presenting a poster.

14.1 What Is a Poster?

A poster is primarily a visual platform for scientific communication. The presentation of a poster is designed to be used in the conference or congress setting as opposed to the oral presentation, which also has a place at smaller meetings and public lectures. The poster shall convey the message of your project, your results and conclusions in an abbreviated form, strongly based on visual means. During a conference you are often required to stand next to your poster for a very specific time frame, the so-called **Poster Session**. Here you are often expected to give a short 2–5 min presentation of your work and to be ready to discuss your poster more in depth with other researchers attending the conference. For the rest of the conference the poster will often be hanging for your fellow delegates to study without you being present. It must, therefore, be self-explanatory, so that everyone can understand your message even if you are not there to guide them.

As you may imagine, composing a poster is a challenging task. On the other hand, if handled well, it has some unique advantageous features attached. First of all it forces you to present your work in a very brief format, which comes in handy on other occasions. It also allows you to have one-to-one contact with interested colleagues and establish strong contacts. In the following we will walk you through the process of making a poster.

14.2 How NOT to Design a Poster

There are many dreadful posters out there presumably owing to the fact that researchers in general spend too little time preparing them. Preparing a poster, especially for the first time takes time, a lot of time! So start early.

We will begin by highlighting some common mistakes that on occasion inappropriately find their way onto posters.

- Too much text. A poster is not a reprint of your manuscript. Write only what is absolutely necessary to understand your message.
- Too many messages. It is impossible to convey several messages properly on a poster so choose your main message and convey it clearly.
- Too few images. Images and figures are a great way to convey your message. An image tells a thousand words. Utilise this fact in your poster.
- Low-resolution images. This is simply a no go. If your figure has too low resolution to be magnified, redo the figure or do not use it. See Figure 14.1.
- Skewed images. Skewed or distorted images that have had relative height and width altered to fit a specific area on the poster looks unprofessional and should most certainly be avoided. See Figure 14.1.

14.3 How to Make a Poster

Having dealt with some of the most common mistakes now is a good time to provide you with some important tools for producing a visually attractive poster that also serves its purpose.

There are two ways to design a poster: the easy way or the hard way. The easy way involves sending your text pieces and figures to a professional graphic

Fig. 14.1 The right panel shows how you should NOT present your figures. Always make sure your figures are high resolution so they do not blur when printed. Also you should never skew your figures to make them fit a specific area

artist who then constructs your poster for you. This often results in very nice posters, but it tends to be expensive and you often need to start the process well in advance to be sure to meet your deadline. Hence, you will in most cases be better off with the hard way of materialising your poster yourself.

There is an infinite number of ways to compose a good-looking poster. The only limitation is your creativity. Start out by choosing an appropriate piece of software for making your poster layout. There are a number of different solutions available for the purpose. We have chosen to abstain from mentioning specific software, but make sure you choose a piece of software that is manufactured for the purpose of producing graphics. If in doubt ask around in your group which programme your colleagues routinely use. Starting with that choice, you ensure that help is at hand if you encounter any problems.

As mentioned above take a close look at the poster guidelines provided on the conference website. Sometimes they may have specific guidelines other than the dimensions of the poster. If so, stick to the guidelines! But usually there will be a lot of room for personal creativity; we encourage you to use this opportunity to create an attractive poster that stands out from the crowd.

There are, however, some mandatory contents that should always be included in your design. First of all your poster must have a title. You have most likely already provided that to the conference organisers when you submitted your abstract. Stick to the same title even though you might have thought of a better one in the meantime. The organising committee will publish a scientific program that will contain your submitted abstract and will provide a reference number for your poster corresponding to its location in the poster hall. If colleagues have read your abstract in advance and would like to know more, they will be looking for your poster based on the title of your submitted abstract. Below the title include the list of study participants and their institutional affiliations. Make sure you highlight yourself preferably by a portrait photo and your contact details. This shows that there is a person behind the science and you may even be recognised in and around the conference venue and thereby you improve your networking opportunities. Finally, you should add logos of your institutions somewhere on your poster.

The poster can vary greatly in size depending on the conference in question, so pay close attention to the guidelines provided on the conference website. It is important to adhere strictly to the dimensions given in the guidelines because you are usually allocated a very specific area for your poster, which you are not allowed to exceed.

14.3.1 The Contents of Your Poster

You can preferably consider the IMRAD structure (Introduction, Methods, Results And Discussion) that we have touched upon earlier as a template for the contents. Your poster should explain the background of your study, the methods used, the results and the discussion and conclusions. Take your point of origin from the abstract you submitted to the conference, shorten it to an absolute minimum while emphasising the most important points. When you have selected your text consider if anything can be shown in the form of illustrations instead of text. Try to avoid tables and use plots and graphs instead if you can. Also consider which colours to use. Seek inspiration in the figures you intend to show. Select colours that are already present within the figures. This introduces a nice harmony to the poster. Make sure your colours look nice together and do not use too many colours. Keep it simple. If you are not comfortable combining colours seek inspiration on the Internet. There are several websites that can help you make nice colour schemes. Again you should keep in mind that high level of contrast between the different elements will ease reading from the distance.

Your poster should be easy to read and have a clear order and flow. Help your reader by attempting to establish a visual pathway, which guides the eyes of the reader logically through the poster. In Figure 14.2 the white path with roman numerals does this, but there are an unlimited number of other ways to achieve this. Highlight the most important parts such as conclusions or important perspectives, but do not highlight more than absolutely necessary.

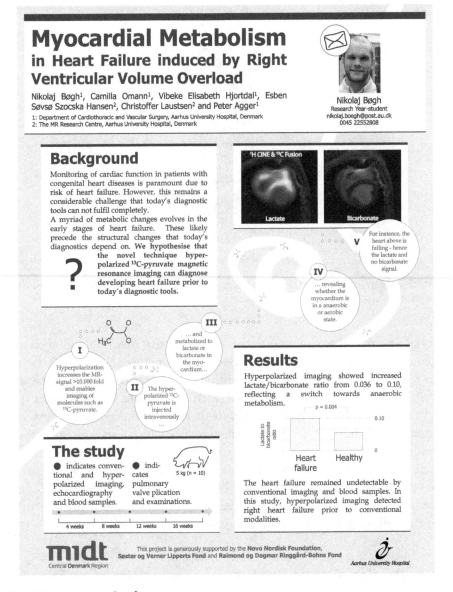

Fig. 14.2 An example of a poster

14.3.2 Attracting Attention

Imagine a poster exhibition at a conference with hundreds of posters hanging side-by-side. Your poster will be just one among many others; so you will need

Contrast Rich

Moving

Big

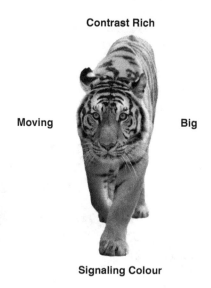

Signaling Colour

Fig. 14.3   Means to attract attention to a graphical presentation such as a poster or a slide show. *Image source: Shutterstock.com*

to consider how to attract attention and awaken the attention and curiosity of passing colleagues. As a consequence of human nature and instincts dating back from the times where we were hunting mammoths with spears, there are certain things that will tend to attract attention. Figure 14.3 illustrates this point, the image shows signalling colours such as red, orange and yellow, big contrast rich elements, and moving elements, all of which attract our attention. Obviously, it is not easy to make a poster move, but the remaining concepts can be easily adapted into a poster. Use less text and more graphics, and make sure your most important elements are big enough and that your font sizes are not too small. Make a test print of a part of your poster in real size and stick it to the wall. All texts should be easily readable from 2 to 3 m away.

14.4 Printing Your Poster

When you have finished designing your poster you should obviously get it printed. Normally you would choose a professional provider for this task or maybe your institution has a provider of its own that you should use. Traditionally, posters are printed on paper, which you then have to transport to the conference nicely rolled in a poster tube. Transporting your poster over longer distances on various flights can be cumbersome, because it is advisable to carry your poster as hand luggage. Never let your poster be transported as conventional luggage. Poster rolls actually do roll! They easily fall off conveyer belts and will often not be found until your plane has taken off. Imagine arriving to the conference without your poster. Then you have a serious problem! An alternative to the paper poster is to have the poster printed on fabric. They generally do not look quite as nice as the paper versions, but on the other hand they fold nicely into a carry-on bag or suitcase and are much easier to transport.

14.5 The Electronic Poster

As a relatively new concept, at some conferences you can bring an electronic poster. Essentially it is exactly the same as a standard poster, but in this case it is not printed. It will be shown at the conference on a large monitor. Bear in mind that monitors cannot be expected to have the same resolution as the printed poster. In this case you need, therefore, to pay even more attention to the size of your figures and texts. In particular thin lines in plots can be very difficult to distinguish and figure legends can be very difficult to read.

14.6 How to Present a Poster

Now for the important bit; the presentation of a poster. This is probably the most difficult presentation there is. This is not said to scare you away from doing it, but to encourage you to practice your presentation like you have never practiced before! The reason why this particular presentation is challenging relates to the very limited time frame available. Normally a poster presentation is restricted to a few minutes. It is very important that you are aware of exactly how much time you will be allotted, and that you stick to it. The less time you have, the more you have to prepare.

If you want me to speak for two minutes, it will take me three weeks of preparation. If you want me to speak for thirty minutes, it will take me a week to prepare. If you want me to speak for an hour, I am ready now.
- Winston Churchill

Always start out by introducing yourself and your institution. The main presentation should be structured in the usual way, namely the IMRAD structure. But because you have limited time you should choose only one main message, so choose the most important one. Less is more! There will be many posters to look at, and your colleagues will, therefore, struggle to remember more than one message. Do not overcomplicate things. After your presentation there will usually be allocated time for questions and discussion; this gives you a chance to elaborate and to provide further information. We will cover the scientific discussion in the following chapter on oral presentation. At some conferences you will not be asked to prepare a presentation as such, but you will instead be given a time slot where you are expected to be standing beside your poster. Colleagues will then pass by and ask questions and engage in discussion as they see fit. Also in this situation, you should prepare a quick presentation of your work designed to be given one-on-one according to the same guidelines as above. The "Elevator pitch" described in Chapter 19 can be of good use here. Consider bringing some A4 prints of your poster that people can take home.

Presenting a poster is not an easy task, but it is great fun. It is a unique opportunity to interact with your colleagues face-to-face, get feedback on your work and do some networking. You can learn more about effective scientific networking in Chapter 19.

15

The Oral Presentation

The oral presentation is the hallmark of scientific communication. Standing in front of an audience talking about your scientific endeavours is one thing, but holding the audience's attention in the palm of your hand while still delivering your message in a captivating manner is a different thing entirely. That said delivering a powerful and memorable oral presentation is not that difficult; it does, however, require that you focus a little less on your slide show and more on the presentation itself. Often people think of an oral presentation as being just a slide show, but they could not be more wrong. The slide show is a visual aid that can help you convey your message in an inspiring way, but the presentation is in fact …you! If you do not accept this role you will be unlikely to succeed. Do not try to hide. You are the presentation!

We cannot possibly cover all the aspects that contribute to the "perfect" oral presentation, but we will try to sum up the most relevant points to help you as a new researcher prepare for your first talks. There is much to learn in this field, not only for the novice, but also for the experienced researcher.

Often you would think of an oral presentation in a formal setting where a scientific message is verbally passed on to an audience. But the oral presentation belongs not only in the conference setting, but also in a university lecture hall or a casual morning meeting at your department. The hints and tips given in the following chapter apply to any setting where you are required to present information to your colleagues or peers.

© Springer International Publishing AG 2017
P. Agger et al., *A Practical Guide to Biomedical Research*,
DOI 10.1007/978-3-319-63582-8_15

15.1 Preparing Your Presentation

When preparing for an oral presentation at a conference, the first thing you should do is find the presentation guidelines on the conference website or in the programme. If you are in any doubt, contact a representative of the conference by email. Specifically, pay close attention to the time allocated for your talk. How much time will be allocated for the talk itself? And how much for discussion and questions? The time frame for an oral presentation is to be taken very seriously! We cannot stress this enough. There are no valid reasons for running overtime. Not only does it convey a blatant disregard for your fellow speakers, but there is nothing more annoying for the audience. The second you exceed your given time frame your audience will start looking at their watches in anticipation of your ending, and you will progressively lose the attention of everyone in the room. Running overtime is NOT an option!

> Death (or worse) to those who run overtime!
> - Unknown

When consulting the presentation guidelines you should also note if there are any requirements in terms of visual aids. What equipment will be available for you to use? There are many types of visual aids that you can make good use of if available, such as blackboards, whiteboards, flip charts and so on, but normally you will have to make do with a slide show projected onto a large screen. Note that you may be forced to use a specific slide show software, because some conferences will insist that all presentations are delivered using the same central conference computer. On other occasions you may be allowed to use your own computer. Make sure you know what to expect.

15.2 Composing a Slide Show

The first thing you should do when preparing a slide show is step away from your computer, find a pad and a pencil and create an outline of your talk. This may seem counterintuitive, but it makes good sense, try it! Think of your talk as a storyline, and plan how to direct your audience through its main messages. Doing so allows you to think freely and outside the box, away from the constrictions of the computer. It, furthermore, prevents you from giving in to the temptation of just building your presentation without a clear idea of the direction and goal of your talk. Always keep your target group in mind

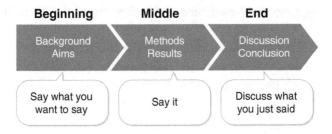

Fig. 15.1 A presentation has three sections, the beginning, the middle and the ending. The beginning contains the background and the aim of your study. The middle contains the methods and the results, while the ending contains the discussion and the conclusion. This gives you the opportunity to almost say the same things three times

and keep your language appropriate and make sure to pass on your message clearly. Keep it as simple as possible.

Do not expect too much of your audience. You want them to remember your talk, but there is a limit as to how much one can be expected to remember. Therefore, choose a maximum of three main messages for your talk. We know you probably have much more to say, but leave that for now. When it comes to effective delivery of a scientific message less is more.

Construct your talk into three main parts: the beginning, the middle and the ending (Figure 15.1). In each part you should approach your main messages from three different angles. In the beginning you give the background for your study and finish with your aim. In short you tell the audience what you are about to tell them. This section should grab the audience's attention. It also gives you the opportunity to get rid of any underlying nerves or anxiety. In the middle section you provide an overview of your methods, and the most important results relative to the three main messages of your talk. People tend to remember the last thing they heard. Utilise this in the final section— the ending. Discuss your results relative to other studies and provide your conclusions by repeating your three main messages and putting them into perspective. In short, discuss what you just said.

When you have an idea of how your slide show will be constructed, start putting it together on your computer. There are many kinds of software for presenting, and no one is particularly better than the other. Chose whichever you like the best, or the one that is commonly used in your department. As we discussed in Chapter 10 on poster presentation, you can steer the attention of the audience by the use of contrast, size, colour and animation. Be careful though with animations, they can be helpful if used in moderation, but if used too extensively they can be confusing and highly disturbing. If in doubt, leave it out!

Use the slide show as a visual aid. Your slides should support your talk, not be your talk. Reduce the amount of text to an absolute minimum and most importantly, if you are showing a slide with a lot of text on it, do not talk while showing it. People cannot read and listen at the same time.

> People cannot read and listen at the same time!
> - David Philips

Try to reduce the number of messages per slide. Preferably one slide should convey one message. You can always split slides into several slides if you have many massages to deliver. On top of that you should clean up your slides and get rid of everything that does not serve a specific purpose. You will often see that researchers fill up their slides with objects, icons and text that are not needed to deliver the message. The human mind uses a lot of energy interpreting what it sees. Why force your audience to spend energy on interpreting something if it is unnecessary?

15.3 Arrival at the Conference Venue

Upon arrival at the conference, you will have to investigate what you should do with your slide show. Often larger conferences have a central server to which your slide show should be uploaded in due time before your talk. At smaller conferences you may be expected to bring your slide show to the room designated for your talk before a certain time to load it onto the local computer. Make sure you know what to do. If possible go see some oral presentations before you do your own. Presentation settings vary slightly from conference to conference and it is always good to get a feeling of what to expect.

15.4 On the Day of Your Presentation

Arrive in good time to get a feeling of the setting. In some settings you may have the opportunity to test your slide show to see if everything works as intended. If you get the opportunity, use it! Especially videos and animations tend to have a will of their own, it is always good to check they are behaving!

Source: Shutterstock

15.5 Giving Your Presentation

Start out by expressing your gratitude to the organising committee and the chairmen for being granted the opportunity to present your work. The chairman will usually introduce you. If that is the case you do not have to do it as well, you will just loose the interest of the audience by repeating what has been already said. If the chairman fails to do it, introduce yourself with name and affiliation. In favour of keeping time, the moderators are allowed to stop you or request that you finish your talk quickly if you run overtime. Such interruption is not a pleasant feeling, so you better make sure this does not happen. We have stressed the importance of keeping to time, but this does not mean you should talk as quickly as possible to fit everything in! Remember less is more. Often conferences are international events where English is not every delegates first language, speak at an appropriate speed, ask your colleagues for feedback on this matter after your practice talks.

15.5.1 Controlling Anxiety

It is absolutely okay to be nervous when presenting. Most presenters, even the most experienced ones, struggle with anxiety to some extent and that is good. Anxiety is a natural adrenalin kick that helps you focus, but it should not take the control out of your hands. Taking control of anxiety can be a difficult task, but in general you gain control by convincing yourself that you know what you are doing. This comes with practice. The more you practice your talk, the more confident you get. Give your talk in front of colleagues, several times if needed. Learn the first 30 s of your talk by heart, so you do not have to spend energy remembering what to say. In that way you do not have to worry about forgetting important things in your introduction and you can allow yourself to focus on relaxing. If you have problems with controlling your anxiety try to get on stage before your audience arrives to get a feeling of the room.

It is a matter of style and preference whether you should bring a script to the podium, but in general, a presenter who just reads a pre-written script aloud is not as inspiring as a presenter speaking organically and connecting directly with the audience. Our advice is to avoid the use of a script, it is very restrictive; but if you feel happier having a few keywords as a support, it may be a good psychologic support for your initial talks.

> Don't practice until you get it right. Practice until you can't get it wrong!
> - Unknown

15.5.2 Body Language

Conveying a message successfully largely relies on trust. If you do not appear trustworthy when giving your presentation, your audience will not remember what you said. Trust is of course conveyed through reasonable arguments, but approximately 80% of all communication between humans happens without words—through body language. So if you want to appear trustworthy consideration of your body language is a good place to start.

When thinking about body language in this setting you have to consider four things. Your eyes, hands, posture and legs (Figure 15.2). Look primarily at the audience, not at your slides, and if possible occasionally look people directly in the eyes. Maybe you are not comfortable with that, so in that case just look at peoples' foreheads or the wall just behind the back row. It does not work quite as well, but nearly. Make sure that you know what you do with your hands. Do not cross your arms in front of you, that signals arrogance and negligence to the audience. The same is the case if you put your hands in your pockets, or hide your hands behind your back. Put your hands where everybody can see them. Use them for relevant gestures, but also remember that just letting your arms hang casually along your sides on occasion does not look as awkward as it feels.

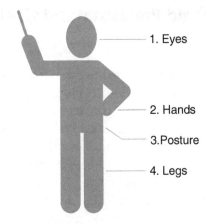

Fig. 15.2 Your all important focus points when considering your body language when presenting

> Put your hands where everybody can see them.

Never turn your back on the audience. Always position yourself so the front of your body is directed towards the audience. This is what we mean when telling you to focus on your posture. In that way you signal that you are willing to stand up for your results and arguments, and have no intentions of hiding anything. If you appear fascinated and passionate about your work, you will capture your audience.

Lastly, consider the whereabouts of your legs. Often you have been granted a whole stage for your presentation, by all means use it! Moving on stage allows you to address different parts of the audience making everybody feel included. A bit of movement on the stage also helps you to maintain the attention of your audience. You will, however, sometimes find yourself in a situation where you are restricted to stand behind a speaker's stand or podium, maybe because your microphone is fixed there. This is in no way a preferable situation, because it impedes your possibility to create dynamic movements during your talk. Unfortunately you cannot control everything and sometimes you will just have to make the best of a bad situation, remember the conditions are the same for everybody else.

15.6 The Handheld Presenter and the Laser Pointer

At most conferences you will be allowed to use a presenter, which is a wireless device connected to your computer that allows you to go back and forth through your slides. This is a very convenient device because it releases you from the confines of your computer. Preferably, you should buy one for yourself so you always have one on hand that works and that you are comfortable with. Most presenters also house a laser pointer. Also known as the distractor! Contrary to what you may think, this is something that you do not need! If you think you need a pointing device to guide the attention of your audience, your slides are probably too complex and should preferably be split up or in other ways simplified. On top of that a laser pointer magnifies just the slightest shaking of your hands—so if you are a bit nervous during your presentation, and sometimes even if you are not, a vibrating red laser dot on the screen does not help either the audience or yourself to focus. Forget to take your finger of the button …and we now have a weapon capable of blinding your audience members one by one with laser precision!

15.7 Oh s…! Someone is Asking Questions!

There will always be allotted time for questions, and there will always be someone posing a question after your talk. It is considered impolite not to ask questions so if there are no questions from the audience the chairman will have prepared one, hopefully inspiring further questions. The thought of experienced researchers asking you questions can be horrifying, but remember that most of the time people are not asking questions to hurt you or see you squirm, but merely out of sincere interest.

Try to anticipate potential questions and prepare a good answer for them. Discuss relevant questions with colleagues and supervisors before the conference. When asked a question start out by thanking the individual for the question and preferably point towards the excellent quality of the question. People tend to be friendlier when someone acknowledges the quality of their question. No one expects you to have an answer ready straightaway. What feels like minutes for you on stage, feels like seconds for the audience, so take your time and use a couple of seconds to think of an appropriate answer. Answer the question as well as you can. Be honest and humble, and do not make things up if you do not know the answer. Sometimes "This is a very good question, but I do not know the answer" is the right answer.

15.8 Improving Your Presentation Skills

Through your entire career you should always strive to improve your presentation skills. Ask your colleagues to attend your talk and discuss the good and the bad afterwards. Encourage constructive criticism. A very effective technique to boost your skills is to get a colleague to record a video of your presentation for you to review. This is a forceful technique that is just as effective as it is unpleasant. If you do not like the thought of this ask your colleague to record the video with your own mobile phone. Then you are sure that only you will have access to it. Pay attention to your general appearance, your body language and especially the speed of your speech. You are probably speaking a lot faster than you think.

With meticulous preparation, and engaging presentation, supported by attractive visual material and intelligent handling of discussion, you have the arsenal to take on the essential task that is the oral presentation.

16

The Scientific Manuscript

Research is not research until it is published! At the end of the project life cycle you will condense your discoveries in a scientific manuscript to be published in a scientific journal. As in every aspect of research, when you do something for the first time, it is always a challenge. Composing your first manuscript is no exception. An author's worst enemy is the blank piece of paper. Staring at a sheet of white paper while thinking about the potential extent of what you are going to write can be an intimidating experience. In this chapter we will discuss the different types of manuscripts and then take you through the process of producing a scientific manuscript. We provide you with tools to get started and relieve you from that daunting blank piece of paper.

16.1 Types of Manuscripts

Firstly, you should be aware that there are many different types of scientific papers, each serving an individual purpose. Obviously, the first thing to consider when writing a manuscript is the overall manuscript type. We will, therefore, in the following, describe each type of manuscript, so you have an idea about what to choose between.

16.1.1 The Original Article

This is what is traditionally thought of as a scientific paper. The original article conveys new results and scientific discoveries. Studies described in original

© Springer International Publishing AG 2017
P. Agger et al., *A Practical Guide to Biomedical Research*,
DOI 10.1007/978-3-319-63582-8_16

articles usually include some sort of experiment or data extraction and are drawing conclusions based on new data. It typically follows the IMRAD structure. Because this is by far the most common type of article, we will go through this in detail below in Section 16.3.

16.1.2   Short Communication

Like the original article this form of scientific communication also conveys a message of a new finding. It is, however, considerably shorter. Short communications are often used if you have an important message on a new finding that you need to publish fast, because it has significant scientific value and potential consequences in the research environment, and to some extent also because it is important that you are the first to show this. Usually the publication speed of short communications is deliberately made faster than original articles. To highlight this they are sometimes even called rapid communications.

16.1.3   Technical Note

Like the short communication the technical note is a shorter manuscript, but it is focussing on a new routine or method. If you have invented a new way of doing things or a new device, but have not produced any new results as such, publishing a technical note may be the way to make sure you will get the credit for inventing your method. It is also a good reference document, where you—and others—can refer to the use of that particular technique.

16.1.4   The Systematic Review

Review papers are usually written by researchers with extensive experience within a certain topic. It is summarising and discussing all available literature within a topic. It aims to paint a picture of the overall knowledge within a research topic reaching agreement as to what is known within this field. Reviews are often long and comprehensive; hence they make good references for describing the general view on a specific topic.

16.1.5 Meta-Analysis

In a meta-analysis several clinical trials (see Chapter 2) on the same topic can be pooled together in order to statistically enhance the strength of the conclusions.

16.1.6 Case Report

From time to time rare cases of diseases occur. These can for instance be described in a case report, so that other scientists or clinicians can learn from it. Case studies are normally rather short, just describing the circumstances of the case and typically also the treatment and its outcome.

16.1.7 Video Article

A video article is as the name implies a video typically showing a new approach to a scientific or a medical procedure. They can be accompanied by various amounts of text and usually include an abstract.

16.2 Reporting Guidelines

As can be seen from the above, written scientific communication covers several types of manuscripts, that all differ slightly. In order to improve the quality of the thousands of scientific contributions that are published every year, the different editorial boards have joined forces and completed several sets of guidelines outlining the requirements for different types of studies. In Table 16.1 you can find a list of the different reporting guidelines all named with an acronym. It is a good idea to familiarise yourself with these because they provide a point of origin when you are to write your paper. They help you to remember all necessary details and prevent you from struggling with the blank piece of paper. An example of the CONSORT guidelines is shown in Figure 16.1.

Table 16.1 Overview over available guidelines

Guideline	Description
CONSORT	Randomised trials
STROBE	Observational studies
PRISMA	Systematic reviews
STARD	Diagnostic or prognostic studies
ARRIVE	Animal experimental studies
CARE	Case reports
SRQR	Qualitative research
SQUIRE	Quality improvement studies
CHEERS	Economic evaluations
SPIRIT	Study protocols for clinical trials
AGREE / RIGHT	Clinical practice guidelines

CONSORT 2010 checklist of information to include when reporting a randomised trial*

Section/Topic	Item No	Checklist item
Title and abstract		
	1a	Identification as a randomised trial in the title
	1b	Structured summary of trial design, methods, results, and conclusions (for specific guidance see CONSORT for abstracts)
Introduction		
Background and objectives	2a	Scientific background and explanation of rationale
	2b	Specific objectives or hypotheses
Methods		
Trial design	3a	Description of trial design (such as parallel, factorial) including allocation ratio
	3b	Important changes to methods after trial commencement (such as eligibility criteria), with reasons
Participants	4a	Eligibility criteria for participants
	4b	Settings and locations where the data were collected
Interventions	5	The interventions for each group with sufficient details to allow replication, including how and when they were actually administered
Outcomes	6a	Completely defined pre-specified primary and secondary outcome measures, including how and when they were assessed
	6b	Any changes to trial outcomes after the trial commenced, with reasons
Sample size	7a	How sample size was determined
	7b	When applicable, explanation of any interim analyses and stopping guidelines
Randomisation:		
Sequence generation	8a	Method used to generate the random allocation sequence
	8b	Type of randomisation; details of any restriction (such as blocking and block size)
Allocation concealment mechanism	9	Mechanism used to implement the random allocation sequence (such as sequentially numbered containers), describing any steps taken to conceal the sequence until interventions were assigned
Implementation	10	Who generated the random allocation sequence, who enrolled participants, and who assigned participants to interventions
Blinding	11a	If done, who was blinded after assignment to interventions (for example, participants, care providers, those

CONSORT 2010 checklist   Page 1

Fig. 16.1   First page from the CONSORT guidelines

16.3 The Contents of the Original Article

When you start writing the manuscript for an article, and you already know the journal to which you intend to submit your work, it is the right time to visit the journal website. All scientific journals have a dedicated section in their website entitled "Instructions for Authors" or the like. Here you can

- Title
- Authors
- Key Words
- Abstract
- Introduction
- Materials and Methods
- Results
- Discussion
- Conclusion
- Acknowledgements
- Disclosures
- References
- Table legends
- Figure legends
- Tables
- Figures

Fig. 16.2   The structure of an original article. Sections marked in *red* highlight the core written elements of a manuscript

find a thorough and often very detailed description on how that particular journal wants you to structure your manuscript. You should put a lot of effort into following these guidelines! If you do not they might reject publishing your paper just due to formalities. Following the journal's guidelines from the beginning saves a lot of time later on, because you do not have to reformat your manuscript; a task that can take several days! At the time of writing you do not always know which journal you will submit to. We, therefore, provide some overall guidelines below that can serve as a good starting point. As stated above, the most common type of article is the original article. In the following, we will walk you through the different sections of such a paper. An overview is provided in Figure 16.2.

16.3.1   The Title Page

The first page of your paper is the title page. As the name implies this page contains the title of your study along with other basic information on the manuscript. As we have discussed previously, coining the perfect title is a challenging task. Generally, one should avoid long and convoluted titles, but on the other hand it is important to give the reader an overview of the study and preferably also its key message. The title should attract the reader's attention. Avoid difficult words that are not common knowledge; you do not want to

Title
The Myocardial Architecture changes in Persistent Pulmonary Hypertension of the Newborn in an Ovine Animal Model.

Running title:
Myocardial Architecture Changes in PPHN

Peter Agger*[1,2], Satyan Lakshminrusimha[3], Christoffer Laustsen[2,4], Sylvia Gugino[5], Jesper R. Frandsen[6], Morten Smerup[1], Robert H. Anderson[7], Vibeke Hjortdal[1,2] & Robin H. Steinhorn[8]

1: Dept. of Cardiothoracic and Vascular Surgery, Aarhus University Hospital, Denmark
2: Dept. of Clinical Medicine, Aarhus University Hospital, Denmark
3: Division of Neonatology, State University of New York at Buffalo, Women and Children's Hospital of Buffalo, NY, USA
4: MR Research Centre, Aarhus University Hospital, Denmark.
5: Dept. of Physiology and Biophysics, State University of New York at Buffalo, NY, USA
6: Center for Functionally Integrative Neuroscience, Aarhus University Hospital, Denmark
7: Institute of Genetic Medicine, University of Newcastle, United Kingdom.
8: Division of Neonatology, UC Davis Children's Hospital, Sacramento, CA, USA

*Corresponding Author:
Peter Agger, MD PhD
Dept. of Cardiothoracic and Vascular Surgery
Aarhus University Hospital
Palle Juul-Jensens Boulevard 99
8200 Aarhus N
Denmark
Email: peter.agger@clin.au.dk
Phone: +4578453084, Fax: +4578453074

Financial support:
This work was supported by the Danish Children's Heart Foundation (PA); The Arvid Nilsson Foundation, Denmark (PA); and the National Heart Lung and Blood Institute, USA [NHLBI L54705] (RHS).

Conflict of Interest
None.

Category of study:
Basic science

Fig. 16.3 An example of a title page

lose the reader already at this point. Sometimes you are asked to provide a "Running title" or "Running head" as well, which is an abbreviated version of the title that goes on top of each page in the printed version of the article. Figure 16.3 shows an example of a title page.

16.3.1.1 Authors

Authorship is important. It gives academic credit and can have significant career and maybe even financial implications for the individual. The number of publications is one of the key parameters in the evaluation of the ranking of a scientist. The more high-ranking papers you have published, the stronger your impact on the scientific community. As objective as this may seem, it is a bit unfair, because the number of publications does not necessarily say anything about the scientific quality of the work or your skills as a researcher as such, but none-the-less authorship is a very important measure in the scientific community. So who should be granted an authorship on your article? In general, is it a good idea to settle this very early in the project life cycle to prevent unforeseen disagreements? When writing your protocol (Chapter 8), you should be very clear about who is going to be the first and the last authors, because these are the most prestigious author positions. The first author does most of the work and is often the practical project leader who has conducted the experiment and written the first draft of the paper. The last author, on the other hand, is the senior researcher, the wirepuller who provides the research environment and the final touches to the manuscript. Sometimes the second author is the runner-up in terms of workload, but that does not have to be the case. Ideally you have to have contributed significantly to the work to qualify as a co-author in one or more of the following points:

- Materialisation of the idea and basic concept.
- Data acquisition.
- Data handling.
- Data interpretation.
- First draft of manuscript.
- Editing and writing of the final version of manuscript.

Keep in mind that many journals set a limit as to how many co-authors are allowed on a paper. It is a good idea to consult your supervisor before promising anyone co-authorship to avoid disagreements.

16.3.1.2 Affiliations

Below the title, authors state their individual institutional affiliations. One author can have more than one affiliation. Reference to each affiliation in the author list is most often indicated by superscript numbers. Affiliations are important for the individual institutions, because the financial future of a research institution largely relies on the number of publications published by their affiliated researchers.

16.3.1.3   Corresponding Author

One author must be the designated corresponding author, being the first in line when corresponding with the publisher. Furthermore, the contact details of this author will be printed on the paper itself and used by other researchers who might wish to correspond with the authors at a later stage.

The rest of the title page may vary a bit between journals and may contain keywords, declarations on financial support or conflicts of interest.

16.3.2   Abstract

The abstract for the scientific manuscript is more or less the same as the abstract for a scientific conference as discussed in Chapter 13. The length of an abstract varies somewhat between journals, but is rarely longer than 500 words. There are two overall types of abstracts in scientific manuscripts; the structured and the unstructured abstracts. The structured abstract is by far the most common. It follows the IMRAD structure. Bear in mind, however, that the abstract for the manuscript is a thorough summary that should disclose all your findings and conclusions, whereas an abstract for a conference is more of a sales document for your project as discussed earlier. In 90% of cases other scientists will only read the abstract, so leave nothing important untold and put an effort into writing a good abstract. The unstructured abstract contains the same information as its structured counterpart, but is condensed into one single paragraph with no subheadings. In spite of being the first part of the manuscript the abstract is typically the last piece of the manuscript that is written.

16.4 The Elements of a Scientific Manuscript

As previously mentioned this section describes the nature of an original manuscript since this is the type of manuscript that is most likely to result from your first run through of the project life cycle. The manuscript of an original piece of research typically follows the IMRAD structure; it is thus subdivided into the following main sections: Introduction, Materials and Methods, Results And Discussion. Each section may have subsections that you can define yourself as needed. When you are about to start writing your manuscript you may be well off using your protocol as a starting point. It already contains a lot of information that fits more or less directly into your manuscript. In the following we will go through each section describing their key features.

16.4.1 The Introduction

The introduction of a scientific manuscript does not differ much from that of the protocol. It may be a bit more comprehensive, though. Overall the structure is the same, beginning with a broad description of the field of research in question gradually narrowing down describing the current challenges, how these have previously been assessed and what is missing. The introduction usually ends with a sentence describing the hypothesis and aim of the study. See Figure 8.2 for reference.

16.4.2 Materials and Methods

This section is the cookbook that thoroughly describes every aspect of your experiment, in a level of detail enabling the reader to repeat your experiments if needed. Describe your study subjects that being patients or animals and describe how you recruited and handled them. If you utilised any databases or registries explain the details of your data extraction and if your data was found in existing literature. Also include a passage, if appropriate, on how you conducted your literature search. Explain all your measuring techniques, the equipment you have used and how you used it. Describe how you handled and analysed your data and what software you used.

In a subsection on statistics state the specific statistical analyses you have used. Here you would also explain your power calculations and you would normally also state the statistical significance level you have used in your calculations. Finally, state which software you have used in your statistical analyses. An example of the statistical analyses paragraph is shown in Figure 16.4.

The last section you should include in Materials and Methods is a small section on ethical considerations. This is largely an abbreviated version of the same section in the protocol. Keep it short and concise. Describe all ethical

Statistical Analyses
Normality was confirmed in all primary endpoints using histograms and Q–Q plots. Overall analysis of mural thickness was done in each ventricle by repeated measures ANOVA, while changes in mural architecture was tested between groups using a nested ANOVA model. Results are reported as means followed by SD in parentheses. Post-hoc analyses were done using Student's t-test. All tests presumed a significance level of 5%. Stata Statistical Software, release 11 (StataCorp LP, TX) was used for all statistical analyses.

Fig. 16.4 An example a statistical analyses section in a manuscript

issues regarding patients, experimental animals and the use of registries. If you have obtained sensitive patient data, describe what measures you took to prevent them from falling into the wrong hands.

16.4.3 Results

This section can be a bit challenging to write. Here you present all your results in a neutral, emotionless tone. You are not allowed to interpret your data at this stage at all. Do not state whether your data are good, bad, interesting or irrelevant. Be completely objective. Data should preferably be presented as text or plots and figures. Occasionally it makes good sense to present your data in a table, but most journals encourage you to present your data as figures whenever possible because these are much easier to interpret.

16.4.4 Figures

Nice figures are important. They are the "exhibition window" of your manuscript that encourages the reader to read your work in detail. It takes time and effort to produce good-looking and informative figures. We advise you to build up effective skills in producing good-looking figures from the very beginning. Otherwise you will end up spending lots of time redoing your figures prior to publication. When making a figure always think of image resolution. Keep the resolution as high as possible, preferably at least 300 dpi for photo or image based figures. For plots and other line-art figures choose a format that is resolution independent such as PDF or EPS. Such vector based formats can be enlarged indefinitely without loosing resolution. All journals have specific requirements for figures regarding format and resolution; make sure you follow these guidelines.

You should place your figures at the end of the manuscript or in a separate document for three reasons. Firstly, trying to merge your figures into your manuscript text will most likely cause you considerable headache because most types of text editing software act a bit anarchistic in this matter and mess things up when you edit your text around the figures. Second, the manuscript is much easier to read if you can have all figures available beside the manuscript while reading and third, you will be asked to upload your figures as independent files anyway when you submit. So why bother putting them into your manuscript?

16.4.5 Discussion

In contrast to what you have just read on being neutral in the Results section, the discussion section, allows you to interpret your results and provide subjective opinions on your discoveries. What has your study actually shown? Start out by stating the principle findings. Compare your findings to those of other studies in the same field of research if such studies exist. Then proceed to discuss strengths and weaknesses of your study. Make your study look a little more interesting by dealing with its strengths first followed by its weaknesses. Then you compare the strengths and weaknesses to other studies. This is particularly interesting if your results are different from other studies.

Then you should discuss the perspectives of your study; what is the potential impact for clinicians or other researchers? Does your study change the way you look at the world or your field of science? Lastly, draw the reader's attention towards unanswered questions and potential future research. The final subsection of the discussion is a small paragraph with a conclusion. Together with the title and the abstract this is the most important part of the manuscript. The conclusion is the key message; a short and concise answer to your hypothesis and aim.

> **The structure of a Discussion section:**
>
> 1. Statement of the principal findings.
> 2. Strengths and weaknesses of the study (strengths first).
> 3. Strengths and weaknesses in relation to other studies.

(continued)

4. Impact of the study: possible implications for colleagues.
5. Unanswered questions and future research.
6. Conclusion.

16.4.6 References

Place your references after the discussion. The bibliography must be formatted according to the guidelines of the journal to which you intent to submit. Here it is important that you use a reference handling software as described in Chapter 5. You will not get it right without it.

16.5 Figure Legends

After the references you should insert your tables if you have any, and lastly your figure and table legends. These legends are the small texts that you will find below each figure in the published version of your manuscript, explaining the contents of the figure. Legends must be short and concise, but remember that they should still enable the reader to understand and interpret the figure without the use of the main text, in other words the figures should be self-explanatory. All abbreviations and symbols used in the figure must be explained in its respective figure legend.

16.6 Acknowledgements

After the figure and table legends there are no more mandatory sections you need to remember, but sometimes you may wish to express your gratitude towards colleagues whom have provided help in various ways during the study and in the writing of your manuscript. You can make a dedicated acknowledgement section for this purpose. Often the people mentioned in this section are lab technicians, graphic designers or contributors who did not qualify as co-authors. In this section you may also express your gratitude to foundations, companies or others who have contributed to the work in terms of money, material, equipment, etc.

16.7 Supplemental Material

Often you will find that there is simply not enough space in one scientific manuscript to display all your data. Most journals provide the opportunity to publish supplemental material alongside your manuscript. This material is not printed in the final paper, but can be downloaded from the publisher's website. As such there is no limit as to the nature of the supplemental material, but usually it comprises comprehensive tables or raw datasets. Also you can publish videos supporting your conclusions. Sometimes the journals favour a shortened version of your "Materials and Methods" section for the printed article, but they may then provide you with the opportunity to publish a more comprehensive methods section as supplemental material. You can refer to your supplemental material within your manuscript, so the reader will know that the material is available.

The day when you submit your first manuscript to be considered for publication in an international scientific journal is one to celebrate—with good reason!

17

Scientific Writing

As you have probably guessed from the previous chapters in Part III of this book, an important aspect of scientific communication is effective scientific writing. Being a skilled scientific writer is not necessarily a natural talent, but is a skill that can be acquired. All you need to do is follow a relatively simple set of rules and, of course, practice. This chapter will provide some tips and tricks for writing scientific documents, be it abstracts, poster texts or entire manuscripts. Use this chapter when you need it. That is when you are actually about to write a piece of scientific text.

17.1 The Biggest Misunderstanding

There seems to be a common misunderstanding that if scientific writing is complex and hard to read it is probably good science. Many people seem to feel that the use of complicated language makes you sound intellectual, scholarly and authoritative. Nothing could be further from the truth! The fundamental purpose of any scientific text is to communicate a message. The only thing that matters is that the reader can understand what you are trying to tell them! The best writing is simple and direct. Focus on precision and clarity. Do not alienate your reader!

© Springer International Publishing AG 2017
P. Agger et al., *A Practical Guide to Biomedical Research*,
DOI 10.1007/978-3-319-63582-8_17

17.2 Learn from the Best

There are many ways to learn how to write. The Internet is flooded with guidelines on how to write, but guidelines alone do not make a good author. When reading scientific literature pay attention to the writing style of other scientists, do not be afraid to adopt elements of their style. You will find that this improves your own writing skills tremendously. Writing in science is nothing like writing a novel. There are systems and workarounds that can be applied, practiced and improved. This is what makes the task of writing a scientific paper accessible to all.

17.3 Specific Writing Tricks

17.3.1 Put New Information Last

Normally the purpose of a scientific manuscript is to present novel information or notions. In order to understand new information we often need some old information to tie the new information to. Give the reader some context. If you present new information before providing the relevant context the reader will intuitively go back in the text to find an explanation, and obviously this is not desirable. This is exemplified in Figure 17.1.

✗ Porcine hearts were used, because human heart specimens are difficult to obtain.

✓ Human heart specimens are difficult to obtain, **therefore**, porcine hearts were used.

✗ Researchers have improved the knowledge of kidney stone morphology by using MRI imaging to visualise the kidneys.

✓ Using MRI imaging to visualise the kidneys, researchers have improved the knowledge of kidney stone morphology.

Fig. 17.1 Put new information last. Provide context by presenting old information first. In that way you set the scene for new information and make it easier for the reader to understand. New information is *blue*, old information is *red*

17.3.2 Active Versus Passive Voice

When using the active voice, the subject of the sentence is the person who is actually involved in the action, and the sentence hence tells the reader who is doing what. Conversely, in the passive voice, the reader is told what is done to the subject of the sentence. Try to avoid overuse of the passive voice because it takes away the responsibility of the actions described in the text, which is the wrong message to convey. As a researcher you want to take responsibility for your actions.

In general you should write in the active voice because this makes your text more personal and trustworthy, see Figure 17.2. You are taking ownership of your research. There are, however, a couple of exceptions where the passive voice should be preferred. Use the passive voice when describing objective procedures such as your techniques in the Materials and Methods section, or when you are describing what other studies have contributed in the discussion section. Technical writing and descriptions of processes and procedures are best communicated in this way, see Figure 17.3. A schematic representation of when to use the different voices is presented in Table 17.2.

Passive:

✗ The study was designed to answer the following question.

Active:

✓ We designed the study to answer the following question.

Fig. 17.2 Favour the active voice in the introduction or discussion sections

Passive:

✓ The heart was placed in the scanner.

Active:

✗ We placed the heart in the scanner.

Fig. 17.3 Favour the passive voice in the methods section and consider using it briefly in the results section

17.4 Important Grammatical Notions

If you are a native English speaker most of what follows will not surprise you. But if English is not your mother tongue, you are likely to benefit from these grammatical tips and tricks.

17.4.1 Correct Use of the Tenses

When should you write in the past tense and when should you write in the present? If in doubt the relevant tenses are outlined with examples in Table 17.1. Use the past tense every time you report someone's results, both yours and those of other researchers. Especially remember to write your results section in the past tense (see Table 17.2). Conversely, you should favour the present tense when you discuss results in the introduction or the discussion sections, but here also the present perfect tense can be of good use. You should, however, avoid the use of the so-called non-human agent. For example you should prefer "the authors concluded that…" rather than "the study concluded that…".

Example
"We have shown that oxygen ventilation is beneficial for these patients; Carlson and associates reported the opposite. A possible explanation of this difference is (…)".

Table 17.1 The use of tenses in scientific writing

Tense	Example
Present	– Our method **is** a valid measure of…
	– Their results **prove** the technique to be valid
Present perfect	– This notion **has been** presented by Agger and colleagues
	– Stephenson and colleagues **have suggested** that…
Past	– The specimens **were** placed in the scanner
	– Previously it **was** believed that…

Table 17.2 The use of voices and tenses in a scientific manuscript

	Active	Passive	Present	Present Perfect	Past
Introduction	✓		+++	++	+
Methods		✓			+++
Results	✓	(✓)			+++
Discussion	✓		+++	++	+

Words like **very** and **extremely** are usually unnecessary. Let the readers decide for themselves whether or not your results are "very" important or "extremely" interesting.

Words like "however", "moreover" and "furthermore" can be very useful. They serve the purpose of providing emphasis to your message by changing the direction of your text. Starting a sentence with "however" for example takes away some of that emphasis. You should, therefore, try to avoid using these words as the first word in a sentence. You should, furthermore, not use these words or their synonyms twice in one sentence. Changing the direction of an argument twice in one paragraph will reduce the impact of your point and, moreover, annoy the reader.

Do not use contractions! They belong in the spoken language and not in academic writing. They decrease the reading speed, affect flow and appear colloquial.

"don't" must be "do not", "isn't" must be "is not" and so on.

17.5 Abbreviations

As we have stated previously abbreviations help ONLY the author NOT the reader! So generally you should try your best to avoid them except when reporting units of numerical information; in this case you should remember to put a space between the number and unit, e.g. 203.65 m, 457 mm. On the other hand, when reporting units without a numerical value, you should spell out the unit.

Example
"The thickness was measured in **millimeters**".

17.6 Numbers and Statistics

If you have to start a sentence with a number and unit, both must be spelled out as words. This is something that you should try to avoid if possible because it can get rather confusing. For the same reason, do not start a sentence with numbers greater than ninety-nine.

Example
"**Twenty-nine** subjects were included in our study".

In general numbers below ten should always be spelled out unless you are reporting results.

Example

"We included **eight** piglets in the study. Their mean weight was **9.5 kg**".

If you are communicating a value that is approximately equal to something, use the symbol $\sim$ (tilde). And when you report decimal numbers less than one always report the preceding zero as well, 0.32 is correct, .32 is wrong.

Use the appropriate number of digits: two significant digits for standard deviations; two decimal places for correlations and two significant digits for percentages are usually reasonable.

Example

73 ± 5; $r = 0.45$; $r = 0.08$; 16%; 1.3%; 0.013%.

Obviously, reading this chapter will not in itself make you a good scientific writer. We have chosen only a few relevant remarks and common errors for you to consider, so you are not completely lost when it comes to drafting your first scientific texts. If you are keen to know more, we encourage you to explore the excellent literature available in the field. But please remember that reading alone does not make you a good writer, only writing does!

18

The Process of Publishing Your Research

The idea of publishing your research should be an exciting prospect. Regardless of your position or stature, it is an extremely rewarding and self-enriching achievement for any scientist. You are making your own little piece of history, and you are contributing information that is truly novel, the value of which has been critically assessed by experts in your field. This information will be accessible to anyone in the world from now until the end of time!

Although the process can be quite arduous and frustrating at times, the above passage hopefully provides you with the desire and motivation to publish your research. By doing so, you join the selected group of individuals who have furthered science for the greater good.

In this chapter we will guide you through the process of publishing your research. All the way from choosing the right journal to the day your own little bit of history goes live and your manuscript is published in the scientific literature.

18.1 The Process…

The process of publishing your research is presented as a 10-step procedure in Figure 18.1. Although much of the process will be discussed below, you are encouraged to review this figure before reading on. Do not be tempted to rush through any of the 10 steps, publishing research is a lengthy process regardless of your experience, for instance the review process alone can take anything from weeks to months. Trust us when we say the achievement of finally publishing your work makes it all worthwhile.

© Springer International Publishing AG 2017
P. Agger et al., *A Practical Guide to Biomedical Research*,
DOI 10.1007/978-3-319-63582-8_18

1. Identify what type of manuscript you will produce

⬇

2. Identify a suitable journal

⬇

3. Match your manuscript to the scope and author guidelines

⬇

4. Submission of your manuscript

⬇

5. Editorial consideration

⬇

6. Peer review process

⬇

7. Response to reviewers

⬇

8. Editorial decision

⬇

9. Payment

⬇

10. Publication

Fig. 18.1   The flow of publishing

It is important to realise that you will not always be successful when it comes to publishing your work, it is a cold hard fact that the acceptance of your work by the editor and the assigned reviewers is a relatively subjective process, and you may well have to try multiple journals before your manuscript is accepted. Be resilient and do not lose belief in the importance of your work.

In the case of rejection, suck it up and return back to step one. You are most likely to face rejection at either step 5 (editorial consideration) or step 6 (peer review process). So, if your peers do not deem your work suitable for their journal, do not be disheartened, some of the greatest scientific discoveries and theories in the history of human kind were initially rejected by peers. Find a new journal and begin a new submission process.

Resubmission is always a more efficient process, everything is easier second time around. Furthermore, due to the critical input from the previous review process, it is highly likely that you will actually have improved your manuscript, so that initial rejection can in fact be a positive thing!

Always look on the bright side of life

18.2 Choosing a Journal

Choosing the right journal for your publication should be a considered decision, but importantly you should aim for the best journal possible, do not fear rejection!

> Ambition is the path to success, persistence is the vehicle you arrive in
> - William Eardley IV

Note, the first times you submit a manuscript to a journal it is critical you have an experienced co-worker to consult—most likely this will be your supervisor. They will help judge the impact of your manuscript and advise the right level of journal to target.

18.2.1   What Type of Manuscript will You Produce?

The different types of scientific manuscripts, from short communications all the way to comprehensive reviews are described and discussed in Chapter 16—The Scientific Manuscript. It is important to be clear what type of manuscript you will produce before deciding on a journal. Every journal has different categories, criteria and guidelines for different manuscript types. Furthermore, it may be that the journal you originally wished to publish in does not accept your type of manuscript.

18.2.2   The Journal Scope

The scope of a journal basically outlines the journal's vision and what the editorial board aims to provide for their readers with their publications. The scope of a journal can be broad (e.g. Nature) or very specific (e.g. Journal of Heart Valve Disease), it may concentrate on a specific disease in a specific demographic such as atherosclerosis in humans, or on a broad core subject such as physiology.

You should target the scope of the journal in both your manuscript and cover letter described in Section 18.3.1. For example do not submit an animal study of diabetes to a journal that clearly outlines in their scope, and in previous publications, that they are only concerned with human studies. Equally do not waste your time submitting a manuscript, which investigates a research question using a single technique, when you can see from previous publications they are looking for multidisciplinary studies.

18.2.3 Impact Factor

What is impact factor? As a general rule you should always strive to publish in a journal with the highest impact factor possible, but what does impact factor really mean? The impact factor of a journal for a given year relies on the citations of the papers published in the journal in the two previous years. Citations are simply the number of times a paper has been referenced by other peer-reviewed papers. The impact factor is calculated by dividing **the number of citations in the last year of papers published during the two preceding years by the total number of papers published in the two preceding years.** An example of how to calculate the impact factor of a journal in 2017 is given in Figure 18.2.

An impact factor of 5 means that, on average, the articles published 1 or 2 years ago have been cited 5 times. This gives you an idea of how often publications in this journal are read and the impact fellow researchers assign to information published by this journal. This also gives you an idea of how difficult it is to have your manuscript published in the journal in question. As to maintain their impact factor the editor has to be convinced your manuscript will be cited by your peers. You can also find 5-year impact factors, which is calculated based on the number of citations over a given 5-year period. Bear in mind that the impact factor is very field specific. An impact factor of 5 may be considered high in one field of research and low in another. To get an idea of what is a high impact factor in your field, we advise you to consult your supervisor and collaborators.

How important is impact factor? This is an interesting question and the answer is up for debate. Journals with high impact factors are often well respected and have larger readerships than low impact journals. Publishing your work in such journals, therefore, adds credibility to your work, and inherently improves the exposure and likelihood of your work been read. Furthermore, your peers appreciate the challenge of publishing in high-impact journals. Prospective employers and funders will be looking for the level of journal you publish in, but publishing in high-impact journals comes at both a financial cost and time cost. Journals work very hard to raise and maintain their impact factor, and they know it is important to scientists where their

$$\text{Impact factor } 2017 = \frac{\text{Citations } 2016 + \text{Citations } 2015}{\text{Publications } 2016 + \text{Publications } 2015}$$

Fig. 18.2 Calculation of the impact factor

work is published. As a general rule, the higher the impact factor the higher the financial cost, and the more rigorous and time consuming the submission and review process.

Impact factor is, however, a controversial subject. It can be argued that smaller groups or projects with reduced financial support are priced out of publishing in the high impact journals, regardless of the importance and novelty of their work. An argument can be made that the true impact of a scientific paper should not be judged based on where it is published, but by how many citations it receives. After all, a citation is a recognition from your peers that your work is of importance. Although you will be requested to include the journals in which your work is published when you make a job or grant application, you will often also be asked to provide the number of citations. Therefore, if financial restrictions prevent you from publishing in your desired journal, do not be too disheartened, your peers will also have a say in the impact of your work.

As discussed above, high-impact journals provide increased exposure for your work. With this in mind, when considering lower impact journals you should be aware of their reach. Be aware of journals offering free or low cost publication, they often have low reach, meaning they do not appear on major scientific databases, or they may have low priority in search engines. Also note you cannot resubmit a paper or data once it is published, so if you publish your work in a low-impact factor journal just to make sure it gets published, you cannot then resubmit the same data to different journal with a higher impact factor. Do not work hard on producing novel data just for it to get lost in a shadowy corner of the scientific literature.

Source: Shutterstock

18.2.4   The Financial Cost of Publishing

This is an extremely important point and relates to step 9—Payment (Figure 18.1). Yes, after all your hard work raising funds, conducting your research and writing your publication, you often have to pay for the privilege to publish it! You must, therefore, choose a journal you can financially afford to publish in. Publishing costs can be found on all journal websites.

As we have discussed previously, you should always consider the cost of publishing when planning the budget for your protocol (Chapter 8). You should also be aware of any additional funding, beside your own, which your department may award for publications. In addition, as with authorship, you should have an understanding with your collaborators as to how the cost of publishing the manuscript will be covered. Imagine having your manuscript accepted for publication only to fall down at step 9 of 10 (Figure 18.1). Do not let this happen!

You may have noticed while conducting your literature search that some journals require subscriptions while others are "open access". The price of publishing varies considerably between journals, but with open access you guarantee anyone anywhere can read your work. There is, however, always a trade-off between the impact factor of a journal, its level of access and the cost of publication. All have value to be considered when deciding which is worth the most to you.

18.2.5   Guidelines for Authors

On the website of every journal you will be able to find a link to the guidelines or instructions for authors. As the name suggests this provides you with information on how to compose your manuscript, along with limitations such as maximum word count, number of figures and number of authors. It is, therefore, important to review these when choosing a journal. In general you should always adhere to the guidelines, but in certain situations the better solution is to choose another more appropriate journal. Maybe the guidelines are too restrictive to accommodate the full scale of your scientific message. For example, you should avoid trying to truncate large multidisciplinary study into say 2000 words, and you do not want to get to the point of submission just to realise you have exceeded the number of figures, and additionally have too many authors. There are although a few tricks, which allow you to avoid such problems. As briefly described in Chapter 16, you can add extra information such as detailed methodologies and additional figures as supplemental material, which is often free from restrictions. Just make sure you reference the data in the main manuscript! Furthermore, you may be able to negotiate some of your co-author's authorship down to an acknowledgement if their contribution was limited.

18.3 Submission

Before initiating the submission process, it is good etiquette to send the final version of the manuscript to all of your co-authors. You should thank them for their input, and provide them with a time frame (perhaps 3–5 days) to make any comments or amendments prior to submission. The submission process is an online process and you must make sure that you have complied with the guidelines for authors. Author guidelines are different for every single journal, and you should always meticulously adhere to them. Glaring ignorance to journal guidelines appears sloppy and unprofessional. Due to the variability between journals, we will not provide an all-inclusive list here, but the main points to be aware of are:

- Manuscript layout (e.g. IMRAD).
- Word count.
- Number of figures and formatting.
- Number of authors.
- Referencing style.

You are advised to use referencing software, not only does it avoid errors, it also gives you the opportunity to download the specific journals referencing style, which you can then apply to your document.

18.3.1 The Cover Letter

You should always include a cover letter with your submission, an example is given in Figure 18.3. The cover letter is not crucial to the acceptance of your manuscript, but it is a formality you should respect. It allows you to make a personal connection with the editor. Your cover letter should always be addressed to the editor in question, and should portray your enthusiasm to publish in their journal. It should be clear why you wish to publish in their journal, target this comment to the journal scope. Include the title, main

Dear Editor.

We hereby ask you to consider the manuscript entitled *Changes in Myocardial Architecture in an Ovine Animal Model of Persistent Pulmonary Hypertension of the Newborn as assessed with Diffusion Tensor Magnetic Resonance Imaging* for publication in *Pediatric Research* as a *Regular Article*. Neither the manuscript, nor any part of it, has been published previously, nor is currently under consideration for publication by any other journal. All co-authors have approved this submission with no conflicts of interest.

We investigated the changes in myocardial architecture brought upon by persistent pulmonary hypertension in newborn sheep. The novelty in our contribution lies in the combination of novel techniques for imaging the ventricular mural architecture, along with use of an elegant yet complex animal model of persistent pulmonary hypertension in the newborn. We provide evidence that the myocardial architecture is indeed changed due to pulmonary hypertension. Furthermore, we have found a novel anatomical arrangement of the right ventricular mural myocardium.

We hope you will find our work suitable for publication in your journal.

Sincerely yours,
Peter Agger, MD PhD
Aarhus University Hospital
Denmark

Fig. 18.3 Example of a cover letter to accompany a manuscript in a submission

novelty of the paper, and highlight its potential impact, but be concise, your cover letter should not exceed half a page. State what type of paper you wish to submit, for example, original article or short communication, etc. Finish your letter by thanking them for considering your work for publication and state any conflicts of interest.

18.4 Editorial Consideration

Once received by the journal, the editor will first consider whether it matches the scope of the journal, and whether the information presented is of interest to the readership. At this stage your manuscript can either be rejected or sent for peer-review. At this point no news is good news, because rejections at this stage usually find their way to your mailbox rather swiftly.

18.5 The Review Process

The review of your manuscript is conducted by peers, which the journal has deemed of sufficient experience and expertise to provide an informed critical analysis of your submission. The review process is often conducted by 2–5 individuals and can be a lengthy process, taking anything from a few weeks to a few months. Although the reviewers may compliment your work, the main purpose of the review process is to find flaws and errors in your work, and critically analyse the validity of your scientific message. You should, therefore, be prepared to defend your work. It is common for reviewers to be critical, they are not being mean, they are simply doing their job! View their comments objectively and use them to improve your manuscript. Regardless of how well you have prepared your manuscript, you can always make it a little bit better.

It is worth having an idea of what kind of questions the journal will ask the reviewers during the review process; here are some common questions:

- Does the manuscript match the scope of the journal?
- Will the manuscript be of interest to our readership?
- Does the manuscript offer novel information?
- What is the likely impact of this study?
- Does the aim, hypothesis and conclusion match?
- Is the written quality of the paper of sufficient standard?

(continued)

- Have relevant methods been described and applied?
- Are the conclusion supported by the results or have speculation been made?
- Do you suggest any amendments?
- Overall, do you think this manuscript should be published?

18.5.1   The Editor's Response

After the review of your manuscript has been completed, each reviewer report will be forwarded back to the editor. He or she will now make a decision as to whether the manuscript should be rejected or whether you as the author should be given the opportunity to respond to the reviewers' comments. If your manuscript is rejected at this stage, graciously accept the decision, request to be forwarded the reviewers' comments and return to step 2 (Figure 18.1). If the reviewers' comments are sufficiently favourable the editor will provide you with the opportunity to make a response.

18.5.2   Your Response

Always remember the rigours of the peer-review process are essential to maintaining the quality of published science, think of it as quality assurance. Remember your reviewers are not getting paid for the pleasure, so respect this regardless of how critical they are, and keep your responses polite, civil and professional. Creating feuds with other scientists only slows down the progression of science, let your results do the talking!

No matter how frustrating and unfair a review response may appear, take 1–2 days to digest the comments. Sleep on it, the comments always appear more appropriate and less unfriendly once you have had chance to reflect. As a rule, never write your response to reviewers the same day you receive it!

Start your response by thanking your reviewers for their efforts and excellent questions. Go through the reviewer comments one-by-one. State the specific comment followed by your response and the specific changes you have made, see Figure 18.4. When it comes to amendments pick your battles, "don't sweat the small stuff", make any small or minor linguistic or reference changes without kicking up a fuss. You may find you disagree strongly with some comments, in this case formulate a well-founded and referenced counter argument, their comment is not the gospel, one of the joys of science is the

```
Minor comment 1:
Line 41: Replace "3D"  with three dimensioanal.
```

Response 1:
Now amended

```
Major comment 2:
Line 24: explain that myocytes are not spheri-
cal, and that they tend to form aggregates with
a clear preferred direction.
```

Response 2:
Thank you for directing our attention towards the need for this elabora-
tion. We have added the following in the text:
*Page 4, line 2: (...) direction of the cardiomyocytes, being elongated
cells forming aggregates with a preferred direction.*

Fig. 18.4 Example of how to respond to reviewer's comments

right to debate. Ultimately, you may convince the editor, in this case the reviewers comment will be disregarded.

Remember to carefully proof read your response and edits prior to re-submission, and let your co-authors to do the same. You will be given a deadline for your response. Make sure you abide by it, late submission will not be looked upon favourable and the editor may question your desire to publish in their journal. Even though you have been asked to revise your manuscript there are no guarantees; your manuscript can still be rejected at this late stage, it all depends on the **editorial decision** (step 8, Figure 18.1).

If accepted, you will be asked to review a proof copy of the final layout of your manuscript. Often you are only given a few days to complete this review and inform them of any amendments. It is good etiquette to forward the proof to all co-authors and notify them of the deadline for your review.

18.6 Publication

Step 10 of 10......publication. This is a huge achievement for any scientist, make sure you take time to enjoy it! You have now made your own little bit of history, and have made a truly novel contribution to science. This information will be accessible to anyone in the world from now until the end of time! Congratulations! You are a published scientist!

18.7 Public Engagement

How your work can be transmitted to the public for their enrichment is something that is becoming more and more a part of science. Often your department and even funders are keen to see how your work can improve and enrich the lives of the public, be it increasing their understanding of disease states, or providing a hands on experience, for example at public workshops or exhibitions.

Public engagement acts to publicise your work. Social media is a great way to connect with the public and promote your work; you should also contact news companies once your publication has been accepted, often they will be happy to write an online piece highlighting the key points of your work, and they will of course refer to you and your team. Ultimately the impact of your work is measured against the number of citations, and public engagement can provide free advertising for your work. Networking software such as ResearchGate are useful in this setting; they notify you whenever your manuscript receives a new citation, this can be a good way to keep up-to-date with new publications in your field, and may provide you with new networking targets. It is important to be aware that scientific journals often demand exclusivity when publishing your results. Do not make the mistake of publishing your data in other media before publishing in a scientific journal.

You should be aware that public publications are very different to scientific publication. The target group is lay men, therefore the messages have to be accessible to the general public. In the public media it is common to attract attention using exaggerations and embellishments. Be careful not to fall into the trap of over dramatising your research outcome. Early in your career, always consult your supervisor before interacting with the general media.

19

The Scientific Network

A scientific network is basically an all-encompassing term for the group of people you interact with in science. It often includes your supervisors, co-workers, collaborators and associates, and therefore includes people both inside and outside of your immediate circle (Figure 19.1). Your network will continue to evolve and grow throughout your career. Although members may come and go, you should always strive to continually develop and maintain your personal scientific network.

No matter what anybody says, you cannot perform scientific research alone. Even the most famous scientists are a piece of something much larger. There is always a diverse entourage of co-workers and collaborators behind every big shot scientist. Furthermore, in a profession where multidisciplinary investigations are looked upon favourably by funding bodies and scientific journals alike, a proficient and diverse scientific network is vital.

A well-functioning network can provide a wealth of knowledge, inspiration and opportunity. This chapter will highlight the importance of a proliferative scientific network and provide insight into developing, and importantly, maintaining your network.

19.1 The Advantages of Networking

The scientific network is an essential resource in every successful scientist's career. But what is the purpose of networking, how can it specifically help you?

© Springer International Publishing AG 2017
P. Agger et al., *A Practical Guide to Biomedical Research*,
DOI 10.1007/978-3-319-63582-8_19

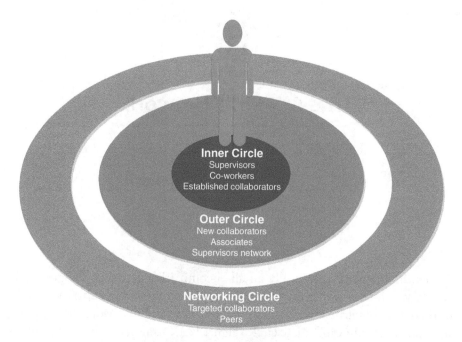

Fig. 19.1 The different parts of your network. The inner circle closest to you comprising your supervisors, co-workers and well-established collaborators. Your outer circle containing newly formed collaborations more peripheral associates and the network of your supervisors. And finally, the networking circle, which contains potential networking targets that you would like to collaborate with in the future

19.1.1 New Opportunities

In "acquiring" your supervisor you have already given yourself access to your first scientific network. In an immediate sense you now have the opportunity to work with or shadow other members of your new lab. But do not hesitate to look beyond your lab. Explore other labs or research groups your supervisor works with at your institution. In a broader sense, you now also have a mutual connection with any groups your supervisor collaborates with, be it national or international. You have a vast range of opportunities open to you. Remember, supervisors will often support researcher exchanges. Why not take advantage of this by working in a new environment or country, learn a new skill or technique, and improve your CV.

19.1.2   Academic Freedom of Speech

Your network by definition will consist of a number of researchers with a wide range of skills and expertise. This means you may often find yourself discussing ideas and concepts beyond your current knowledge. But do not look upon this as a negative or daunting situation, many scientists know a lot about very little, in other words they have a specialism. Use this to your advantage. Your naivety in another field can work in your favour, like an inquisitive child asking their parent questions, your objectivity may well spark a new idea or train of thought. This can be a very enriching and intellectually freeing experience. And do not be afraid of sounding stupid. Your collaborators will happily tell you why your idea is, or is not, suitable. But of course do not try to sound like an expert if you are not. An experienced researcher will see through this immediately. Be curious, be humble and remember there is no such thing as a stupid question.

19.1.3   Improve Your Efficiency

Having a multidisciplinary scientific network, which encompasses your field and beyond can in a number of ways actually make you more efficient. If you take methodological development as an example, this can be a long and labour intensive task. Often collaborative groups will have common research problems they wish to overcome. Keeping up to date with the latest unpublished developments within your network can avoid unnecessary repetition of experiments. Furthermore, you can "share the load". For example, it makes little sense for a busy clinician to learn how to write a mathematical algorithm to model the spread of a malignant cancer, when he already has an expert mathematician within his network, capable of developing such a product in a fraction of the time. Sometimes you need to be realistic, know your limitations and outsource!

19.1.4   Keep Up to Date

Your network will keep you up to date with the latest goings on in your field, often without you even having to lift a finger! For instance, in Chapter 5 we discuss in detail the intricacies of an effective literature search. It is inevitable, however, you will miss the odd important publication. Maybe it has only just been published, or it is still in press. Perhaps, although influential, it was published in an obscure journal, or even originally in a foreign language. Often a collaborator within your network will point you towards such important resources.

19.1.5   Improving Your Future Prospects

"It's not what you know it's who you know", you often hear this quote used in the world of business. Although strictly speaking this does not hold true in the world of science, there is an element of truth to it. The larger your network, the greater your chances of obtaining future research positions, research projects or grants. It is not uncommon in science for researchers to end up working for prolonged periods of time with members of their network, be it a short-term visit, a PhD or postdoctoral position, or even a permanent academic position. The same is true for funding; throughout a scientific career almost all successful grant applications will be the product of collaboration within your network. Some funding bodies only advertise nationally, or require a member of the application to be from a certain country or continent, or are designed solely for national or international collaborative projects. In these instances, a relevant member of your network may open the door to securing a research grant. A diverse network ensures that grant applications and publications are current, multidisciplinary and have impact. This will further improve your chances of future success, as both funding bodies and scientific journals look favourably upon such work. See Chapter 10 for more information regarding applications for funding.

> It's not what you know it's who you know
> - Unknown

19.1.6   Your Scientific Footprint

Promotion of yourself and your work will fuel much of what has been discussed above. Networking offers a tool for enhancing your scientific footprint. Good news travels fast, the larger your network the more people will be talking about you, and the more your name, work and ideas will crop up in the networks of others. When you make a new addition to your network you are not just adding a potential new collaborator, you are gaining access to their network too. Such promotion can aid in expansion of your current network, but importantly raise the profile of your work, increasing the likelihood of those all-important citations. In general, when a scientist recommends a manuscript or technique to another scientist they will go away and look it up. So, do not underestimate the power of "word of mouth" in enhancing your scientific footprint.

19.2 Developing a Scientific Network

We said your supervisor would provide you with access to their network, and this is true, but do not stop there. Eventually you will have to stand on your own two feet in the field, and you will need to build your very own scientific network. Start laying the foundations now!

The prospect of developing a scientific network may appear daunting at first. But there is no rush, it does not happen overnight, it is a continuous process, which will span the life of your career. Here are a few tips on how to get started.

19.2.1 Forming Collaborations

It is important to understand, from the beginning, that you are not obligated to add every researcher you come across to your network. Similar to when you are finding the right supervisor (Chapter 3), you should make sure they are the right fit for you. They should have the potential to enrich your research project, your skill set, or offer future prospects. But remember, a collaboration is a relationship, and like any relationship, it is a two-way street. You should always be able to bring something to the table. This is explored further in Section 19.4 below.

When approaching a potential collaborator you have a couple of options; either you approach them face-to-face in a professional setting, be that at an arranged meeting, or conference; or you approach them in writing, it is advisable to do this electronically using their professional email address. If you are new to networking sometimes the written approach is less intimidating, and has the advantage that your response to questions does not have to be instant. You have time between emails to formulate a considered reply, and potentially research pertinent information or literature to accompany your response. That said you should work hard to develop your interpersonal skills, as networking face-to-face is often far more fruitful. We all make a better connection with people we have actually met. If you ask any experienced researcher, it is highly likely their most productive, fruitful, enjoyable and long-term collaborations will have come about through face-to-face interactions at organised meetings or conferences.

19.2.2   How to Network Effectively

Imagine you are getting into an elevator during a conference. You end up standing beside a colleague you have never met before who naturally asks: "So, what research do you do?" depending on the number of floors you need to ascend, you now have 30–60 s to describe your work. You have the unconditional attention of your colleague up until the bell rings and the elevator doors open. Can you describe your work in 1 min? Trying the "The elevator pitch" for yourself is a highly recommended exercise. The ability to summarise yourself and your work in 1 min is a skill you will use countless times throughout your scientific career, and it is particularly useful in a networking setting. Invest some time in this exercise. Your elevator pitch will need to be adaptable, as a starting point work on producing both a specialist and a lay response, and practice with colleagues and even friends. Make sure you are clear and concise, do not speak too fast, and do not make it overcomplicated. Most importantly you need a "hook", make it interesting, and make an ambitious memorable statement about your work. Remember to be optimistic and passionate; if you do not believe in your work how can you expect anyone else to?

Your goal should be to communicate who you are, what you do, why you are speaking to them and, finally, the "hook"—why they should interact with you.

19.2.3   How Do I Approach a Potential Collaborator?

What is your "in"? Like in everyday life, sparking up conversation with a stranger is not something we practice, in some cultures it may even verge on frankly odd social behaviour. This is, however, what you are met with when networking face-to-face. All you need is an "in", a common interest to get the conversation ball rolling. Use other people, "you know my colleague", or identify a common research problem, or if you are attending a conference, did they or someone from their group give a presentation? If so, compliment them on an interesting talk and have a relevant question prepared. Pre-prepared questions are always a good idea; it prevents the conversation running dry before you have had chance to make your impression. But do not drive the conversation too much, remember a scientific relationship is a two-way street.

Be an active listener, acknowledge what they are saying and appear interested. If they ask you a question, answer it! This may sound obvious but many people will take the opportunity to simply continue promoting themselves, just spluttering information they think is impressive without actually answering the question in hand.

Although talking to a stranger may appear socially challenging, do not forget general social etiquette. A handshake is always a good idea, and of course do not interrupt them in the middle of an existing conversation. Stalk your prey if needed! If you do not get them in the conference hall, look out for them at dinner, in the bar or in and around the conference venue.

Compliments are fine, just do not overdo it, you want to avoid appearing false. But massaging an academics ego with a few compliments relevant to their work and achievements will always go down well. Everyone likes to feel like they have a fan! And do not just talk about work, show them you are a human being, after all, you may end up working together. It is worth showing them you have a personality. Connect on a personal level; discuss the conference food, the state of the coffee, the hotel or perhaps future conference or networking plans they may have.

Source: Shutterstock

19.3 How to Improve Your Chances of Being Networked with

We have already touched upon the fact that networking and collaborations are very much a two-way street. We have discussed how to actively network with potential collaborators, but what about people networking with you. You have just presented some interesting data at a conference, you are now a target for fellow networkers. How can you improve your chances of being networked with?

First and foremost, you want to be memorable, but memorable for the right reasons. Throughout the conference be friendly and look approachable, dress appropriately, do not be the girl in trainers or the guy with the un-ironed shirt! More specific hints and tips regarding this can be found in Chapter 13. You often get the opportunity to answer only 1 or 2 questions after your talk, chances are there will be delegates with unanswered questions looking for an opportunity to chat with you. Therefore, make yourself accessible, and exercise appropriate conference etiquette. Attend as many sessions as you can, and where possible involve yourself in extra-curricular activities. Do not just turn up for your own talk and use the rest of the time to go sightseeing! You may think no one will notice, but trust us they do, again do not be remembered for the wrong reasons.

Advice regarding post conference communication is covered in Section 13.3.3, but there are ways to increase your chances of solidifying new relationships formed during networking. One option is to carry business cards, no they are not exclusively for high flying business types! A simple business card with your name, affiliation and contact details can make all the difference. Not only does it show that you are taking your career seriously, but it also means you avoid awkwardly dictating your name and email, which is always prone to error. When you give out your details, include your chosen name and keywords relevant to the discussion you had, for example, "Robert - cardiac imaging". At the end of a discussion with a particularly promising collaborator it is a good practice to make detailed notes. This means during post-conference communications you do not have to repeat questions or previous discussions. This also serves to improve the overall impression of your competency, dedication and potential as a new collaborator.

19.4 Maintaining a Research Network

Like most things we build, overtime they will require a little maintenance to keep them functioning and running smoothly, and your scientific network is no different. You should be starting to realise your network is a precious commodity, so why not look after it!

Like with any relationship, treat them how you would like to be treated, this is a reciprocal relationship; make sure you are being a good partner. Therefore, you should make sure your collaborations are mutually beneficial. As discussed in the chapter on finding a supervisor (Chapter 3), the best way to ensure to reciprocal relationship is to have clear mutual expectations. For instance, make sure you can handle the work load, be realistic and honest with what your collaborator can expect from you, and do not promise things you simply cannot deliver. When it comes to publication strategy and authorship, decide well in advance, do not do all the work of conducting the study and writing the manuscript just to fall out over authorship. But make sure you are getting a fair deal, even as an inexperienced researcher you should fight for the authorship you deserve. Your supervisor can advise you on this. You can find further advise on this matter in Chapter 16, "The Scientific Manuscript".

Finally, be proactive. Most people involved in research have a passion for their subject, and love the opportunity to discuss science, especially in an informal environment. Organise meetings, seminars or maybe even a small conference. This will only solidify your existing bonds. "If you build it they will come!"

19.5 Data Sharing with Collaborators

You should have reciprocal relationships with the members of your scientific network. Reciprocity literally means "the exchange of things with others for mutual benefit", but how much is too much? Data sharing is a controversial issue, some supervisors will openly share their data, having the viewpoint that science is not a business and should be allowed to progress wherever possible. On the contrary, some supervisors will strictly prevent data sharing of any kind. Make sure you are clear on the sharing policy for your project before you start. In an official sense it might actually be in breach of the study's ethical approval to share sensitive patient data for example. But again make sure you are getting a fair deal. If you share your data you should negotiate what you will get in return, be it data from them, training or authorship on any publication, which may arise.

19.6 Networking Online

There are various online resources, which allow you to build a virtual representation of your scientific network. They are very useful as you receive notifications whenever someone in your network publishes new data, poses a question or begins a new project. Analogies can be made to other non-science social media platforms. You can follow anyone who has a profile, but in this setting your relationship is merely superficial, and should be cemented as described above if you wish to truly reap the benefits. It should be made clear that this is not a substitute for a formally developed scientific network. You are advised to use such mediums merely as a tool to supplement the development and maintenance of your network.

It is advisable to consult your supervisor or colleagues to which resources they find most useful in your specific field. Currently the most pertinent scientific social media platforms are ResearchGate (www.researchgate.net), which is a dedicated research network, and Linkedin (www.linkedin.com), which is a more general professional networking site.

Index

© Springer International Publishing AG 2017
P. Agger et al., *A Practical Guide to Biomedical Research*,
DOI 10.1007/978-3-319-63582-8